The Art and Science of Lecture Demonstration

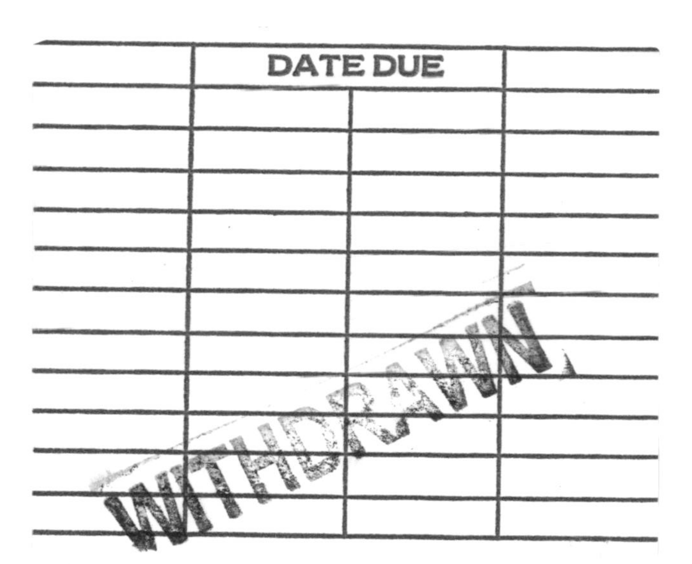

The Art and Science of Lecture Demonstration

Charles Taylor

Professor of Experimental Physics,
The Royal Institution

and

Emeritus Professor of Physics,
University College, Cardiff

Adam Hilger, Bristol and Philadelphia

British Library Cataloguing in Publication Data

Taylor, Charles
The art and science of lecture demonstration.
1. Lecturing
I. Title
371.3′96

ISBN 0-85274-323-8

Library of Congress Cataloging-in-Publication Data

Taylor, Charles Alfred.
The art and science of lecture demonstration / Charles Taylor
196 p. 22 cm.
Bibliography: 4 p.
Includes index.
ISBN 0-85274-323-8 (pbk.)
1. Lectures and lecturing. I. Title.
PN4193.L4T39 1988
808.5′1——dc19 88-11136
CIP

This volume incorporates the substance of
The Gregynog Lectures
delivered at the
University College of Wales, Aberystwyth
during the session 1987–8

Consultant Editor: **Professor A J Meadows**,
University of Loughborough

Published under the Adam Hilger imprint by IOP Publishing Ltd
Techno House, Redcliffe Way, Bristol BS1 6NX, England
242 Cherry Street, Philadelphia, PA 19106, USA

Typeset by Bath Typesetting Ltd, Bath
Printed in Great Britain by J W Arrowsmith Ltd, Bristol

To
My Wife

Contents

Foreword
by
Sir George Porter PRS

This book presents the lifelong experience of a master of the lecture demonstration. It contains a treasury of demonstrations which are described in enough detail for them to be reproduced by others less experienced in the art. Many of them are classics and many others are originals developed by the author and published here for the first time. Lecturers are not usually eager to disclose the secrets of their trade and Professor Charles Taylor is making an invaluable personal gift to his colleagues and to posterity by doing so.

During my 20 years as Director of the Royal Institution I watched hundreds of demonstration lectures and gave many myself. No lecturer surpassed Charles Taylor in the elegance of his demonstrations or the way he enchanted audiences of all ages. His experiences in that great repertory theatre of science and in others all over the world are crystallised in the advice that he gives here on all aspects of his art, from the initial preparation to the inevitable disasters. Never was he confronted by an angry member after a lecture, as was one of his predecessors, with the words 'Young man, those were not Demonstrations, they were Experiments!'

There are very few books on lecture demonstrations and this one is unique in recent times. It begins with a fascinating account of the history of the lecture demonstration, with many references to Michael Faraday and Sir Lawrence Bragg, and Charles Taylor's book will find a worthy place alongside the works of these great pioneers.

Preface

Lecture demonstration has been an abiding passion with me for many years and I was delighted when the invitation to give the Gregynog Lectures at the University College of Wales came along and provided the extra stimulus needed to write this book. There are some other books on the subject in existence, but there is also a strong oral tradition, and many of the ideas included have been gleaned subconsciously over the years and cannot now be traced to their origins. If I have used someone's ideas without due acknowledgment I apologise; my warm thanks go to all those whose suggestions, advice and criticisms have been incorporated without specific acknowledgment, including all the many people who, at the end of one of my lectures, have stayed to chat.

But I should also like to thank many people who have helped in more explicit ways. Sir George Porter, who by inviting me to give a discourse, and subsequently the Christmas Lectures at the Royal Institution, gave me unparalleled opportunities to learn the Art and opened the way to many other developments for which, and for his Foreword to this book, I am enormously grateful. Many other friends and colleagues at the Royal Institution have been most helpful and supportive; I cannot mention them all, but would particularly like to express my gratitude to the present Director, Professor John Thomas for his constant encouragement, to Mrs Irena McCabe, the Librarian, whose genius for tracking down obscure references has been an enormous help and to Mr Bill Coates, who has assisted me in countless lectures, and whose encyclopaedic memory for apparatus and experiments has got me out of a difficulty on many occasions.

My former colleagues at UMIST and at University College, Cardiff, have always encouraged my efforts and helped with useful discussion, criticism and suggestions; but my special thanks go to Professor Henry Lipson at Manchester, who first encouraged me to try my hand at lectures to school-children, and to Mr R S Watkins at Cardiff who has been my mentor in the photographic arts for many years.

Some of the illustrations in the book have been drawn from other publications and I am most grateful to those who have given permission to reproduce them.

The design for the cover incorporates Gilray's cartoon of a lecture at the Royal Institution dated 1802 and is reproduced by courtesy of the Director. Rumford is standing by the door on the right, Davy is holding the bellows, and either Young or Garnett is holding the nose of the Treasurer (Coxe Hippisley).

Charles Taylor

Prologue

When I was a student at the beginning of the Second World War doing a physics course I know that I must have attended lectures on atomic physics, on quantum theory, on electricity and magnetism and on other topics, and I know that some of the lecturers were very distinguished, but I find it very difficult to recall very much about them. On the other hand the lectures on sound (by Dr Alexander Wood) and on physical optics (by Sir Lawrence Bragg) remain as clearly in my mind as if they were given yesterday rather than almost fifty years ago. The reason is not hard to find; they were illustrated by the most superb lecture demonstrations. I am sure that it was through my good fortune in being able to attend these superb examples of the art that I started to use lecture demonstration as one of my main techniques when my career in university teaching began. I suppose I must confess to a certain prejudice in favour of lecture demonstration but, even allowing for that, it seems to me that the evidence for the effectiveness of demonstrations as an information-transfer technique is overwhelming. Sir Lawrence Bragg[1] in referring to the Christmas Lectures to young people and the

> . . .superiority, as judged by the effect on the audience, of a series of experiments and demonstrations explained by a talk over a lecture illustrated by slides

says

> It is surprising how often people in all walks of life own that their interest in science was first aroused by attending one of

> these courses when they were young, and in recalling their impressions they almost invariably say not 'we were told' but 'we were *shown*' this or that.

Lecture demonstration was already popular in the seventeenth century. The Royal Society in 1662 appointed Robert Hooke as curator and demonstrator. G D Bishop[2] records that

> He agreed to provide new experiments for almost every occasion when the Royal Society met, and there was a time when if Hooke was unable to produce an experiment there would be no meeting of the Society.

J T Desaguliers (1683–1744) is often cited as the man who really popularised demonstration lectures and he opens the preface to his *Course of Experimental Philosophy*[3] with the words

> Without Observations and Experiments our natural Philosophy could only be a Science of Terms and an unintelligible jargon.

I shall have a good deal more to say about him in §1.2.

By the mid eighteenth century there were peripatetic lecturers who took round scientific apparatus and who lectured at Eton, Rugby, Westminster, Winchester and elsewhere. Philip Yorke, writing in May 1776, says[4]

> I have been attending experimental lectures this last week and have been much entertained with them. Mr. Brand, who gives the mechanical part, has shown the experiments with good success, for except one or two of the magnetical ones, scarcely any have failed.

But the Golden Age was, I suppose, the second half of the nineteenth century. Science had become a matter of wide public interest; many local scientific, natural history and microscopical societies sprang up and all tended to encourage demonstrations and 'magic lantern' shows at their regular meetings; in the home it became fashionable to perform scientific experiments at table after dinner and many books such as *The Boy's Play Book of Science*[5] (1878) were published. Figure P.1 shows a contemporary woodcut illustrating a typical after-dinner experiment.

Figure P.1 A Victorian after-dinner experiment on coloured shadows using a glass of red wine as a filter. (From *Scientific Mysteries*, 1891, The Chemist & Druggist, 42 Cannon Street, London, p. 80.)

The Friday Evening Discourses at the Royal Institution, started in 1826, succeeded phenomenally because of this public interest, particularly among what were described as the 'professional classes'. The discourses were formal and fashionable, with the lecturer and the audience wearing evening dress, and the atmosphere is summed up by Gwendy Caroe in her delightful informal history of the Royal Institution[6]

> The Managers invited the men who had made the new discoveries to come and talk about it themselves. The audience saw the discoverer and heard him, feeling they were being given first-hand information; felt the thrill on Friday evening when Dewar poured out his liquid air as soon as he could collect enough to pour...

Some questions about the effectiveness of popular science lectures were raised at the end of the nineteenth century. For example, R Galloway[7] wrote in 1881

> I can assure you there is nothing more difficult for a teacher to accomplish than to educate those who have been previously superficially taught. The most difficult to teach I have found, as a rule, are those who have been accustomed to attend popular science lectures; for what they hear is generally very superficial; and, therefore, what they acquire must be *superlatively* superficial.

Hard words indeed, and we shall return to this point in §1.5. But lecture demonstration, both at school and university, remained popular until the Second World War. Why did it then decline? As with most changes, the causes are a mixture of many factors. The growth of scientific knowledge was explosive and more and more material was crammed into the curriculum so that teachers and lecturers tended to feel that it was a waste of valuable time to interrupt their classes by doing experiments; technical assistance was more difficult to obtain; there was less money available to purchase increasingly expensive equipment and in universities the 'publish or perish' principle in determining eligibility for promotion meant that lecturers tended to regard time taken from their research to develop demonstrations as time wasted. Furthermore, many of the more recent discoveries in a subject like physics are impossible to demonstrate because of the scale and cost of the apparatus needed or because of the hazards of the experiment. Sir Lawrence Bragg[8], writing about the duty of a lecturer, blames scientists for the popular ignorance of science and goes on to say

> They are often singularly inept at explaining to non-scientists what they are doing. Further, they are apt to regard colleagues who attempt to give 'popular' talks as actors aiming at popular applause, who cheapen science by over-simplification and spoil the dignity of its aloofness. I am quite out of sympathy with my fellow scientists when they adopt this attitude.

During the last ten years or so the pendulum has begun to swing back again. More and more scientific organisations are putting on Christmas lectures for their local schools;

more and more university departments are offering lectures to schools and, most recently, some organisations are beginning to offer demonstration lectures to primary schools. Why has this change come about? One of the most influential factors has been the decline in the number of students wishing to pursue science and engineering courses in tertiary education, and departments are offering lectures to schools in order to attract more students. The idea of offering lectures to primary schools arose because it was becoming clear that this is the point at which many children are 'turned off' science and at the same time children in the 7–11 age group are full of inquisitiveness about the world around them and are completely uninhibited in their questioning.

In 1985 the Royal Society set up a special committee to study the problem of increasing the public understanding of science. The Report of this committee[9] urged all research scientists to make special efforts to make their work known to the general public and, in one sense, the Report could be said to mark a watershed in that the full approval of the scientific establishment is now given to activities such as the presentation of lecture demonstrations to audiences of children. I am sure that this will give momentum to the movement to restore lecture demonstration to a respected place in the armoury of teaching techniques.

It therefore seemed to be an appropriate time to prepare a book on the art and science of lecture demonstration. My principal purpose is to encourage teachers and lecturers once again to consider the technique seriously and to take the time and trouble to illustrate their lessons and lectures. Since I am a physicist I have confined my illustrations to physics, but most of the principles discussed apply equally well whatever branch of science is being taught.

Of course it is often argued that videotape recording and the video-disc will obviate the need for live demonstrations. Clearly they have their value, as do computer simulations, and these techniques will be discussed in Part 2. Indeed I have taken part myself in the production of video recordings because I think a recording is better than no illustration at all. But, just as the live theatre has a considerable edge over film or television, so I believe does the live demonstration over the 'canned' variety. The principal reason is that the interplay between the lecturer and the audience has a powerful effect. The presentation can be modified to take account of audience response and there is an immediacy and excitement that conveys itself to the audience. I suppose part of this is that there is always the possibility of something going wrong! And this is exactly what happens in real science. I think there is considerable danger in giving audiences polished and perfect productions, such as are commonly put together by the television producers, because they tend to give an unreal picture of what it is like to be a scientist. A nine-year-old girl once complained to me that the reason she didn't like science was that things often went wrong '. . .and when you have to describe something happening that you didn't see it's very difficult'! Her teacher obviously needed some lessons in the art of observation, warts and all!

In Part 1 I propose to spend a little time looking at the historical aspects of the subject. I am not an historian, and I shall not attempt to give an exhaustive or definitive history, but rather to draw out some of the lessons of the past that are still valid today. It is interesting that modern audiences seem to delight in seeing demonstrations using original apparatus. I was recently lecturing in the USA and had taken with me some apparatus from the last century. In particular Wheatstone's 'big glass', that Tyndall used to demonstrate how the vibrations in the musical glasses occurred, excited a lot of interest. One conversation with young members of the audience went like this:

Student: 'Did that really belong to Wheatstone?'
Lecturer: 'Yes'
Student: You mean the 'Wheatstone's Bridge' Wheatstone?
Lecturer: 'That's right'.
Student: 'Can I touch it?'

In Part 1 I shall describe details of some of the famous demonstrations of the past that can still be used very effectively. One of my favourites was first done by John Tyndall in the 1850s. It involved the erection of a thirty-foot-long wooden pole whose lower end stood on the resonance box belonging to a large tuning fork. In Tyndall's manuscript notes[10] the cryptic entry 'Send Barratt upstairs' occurs (see figure P.2). Barratt was, of course, Tyndall's assistant and he went up into the dome, struck the tuning fork with a large mallet and placed it on the upper end of the rod. The sound clearly emerges from the resonator. It is a very rewarding experiment to reproduce because all those present look up to the dome and at the moment the sound emerges from the resonator the whole audience simultaneously looks down at the base of the rod; it is never necessary to ask whether the experiment worked.

17. Transmission through wood: hold rod against tray: tap: touch with tuning fork.
18. Send Barratt upstairs. Long rod from roof resting upon case of tuning fork.
19. Velocity of sound in wood: Tables.
velocity different in different directions: Table
Velocity in metals: — Table.

Figure P.2 Manuscript note by John Tyndall for Lecture 1 of the 1865–6 Christmas Lectures from the archives of the Royal Institution. (Courtesy The Royal Institution.)

Towards the end of Part 1, the whole of which is entitled 'The Growth of the Art', I intend to bring the history up to date with one or two examples of more recent demonstrations, chosen because of their simplicity and elegance, but of course many more current ones will be described in Parts 2 and 3 as illustrations of various aspects of the art and science. Also in Part 1 I shall refer to the presentation of demonstrations on television by the Open University and similar bodies, to the current growth of interactive science museums, which represent a kind of half-way house between demonstration and laboratory work in education, and to the relationship between dramatic presentations of science and the art of demonstration.

Part 2, entitled 'The Science behind the Art', begins with a detailed consideration of the problems of transferring information from the lecturer to the audience. An appreciation of the psychology of the audience and of the lecturer–audience interactions is obviously valuable to lecturers and I have devoted some space to this important question. Part 2 then goes on to consider various kinds of teaching aids which may compete with, or be complementary to, live demonstration, such as slides, tape–slide combinations, closed-circuit television, video recording, video-disc and microcomputers. In particular we shall consider the most useful ways of integrating these techniques into a demonstration lecture.

Part 3, 'The Practice of the Art', gets down to the fine detail of presenting a demonstration lecture, or lesson, beginning with some hints on how to get started, and a survey of various features that are of importance such as care in preparation, choosing the right size for apparatus in relation to the size of the audience and the need for thought about the way in which a demonstration is presented to the audience. Disasters will always happen in even the best-prepared lectures and after a few pages on the problem of coping with such disasters we turn to consider how demonstrations evolve with time, sometimes even as a result of a disaster, or a near disaster.

Taking a lecture on tour outside one's own home ground presents special problems and in this section I have collected together some tips and advice that have been gathered over the last 40 years.

One of the questions that I am asked most often is 'How can you keep the attention of the whole of an audience when they vary in age from 7 to 70, and in experience of science from beginner to expert?' There are some quite simple rules that can be formulated, and after discussing these there is a natural lead in to the question of audience participation and, lastly, to the question of safety.

Finally, in the Epilogue, we round off this collection of reminiscence, advice, recipes and experiences with a discussion of the future of the lecture demonstration and of the all-important need to increase public understanding of science and technology.

Apart from the normal index and the list of references there is an index of demonstrations which also classifies them according to the categories defined in §2.1.

Part 1

The Growth of the Art

1.1 ORIGINS

I suppose the earliest lectures that could be properly be called 'demonstration lectures' are those often illustrated by famous artists of the dissection of a body before an audience of medical students or the occasional instance of a scientist being commanded to demonstrate his latest discoveries before his Monarch. And there are in existence woodcuts purporting to show Pythagoras (figure 1.1) demonstrating some musical phenomena that I suppose come close to being in our category. In the thirteenth century Roger Bacon (1214–92) wrote

> One cannot arrive at the truth except by means of experimentation . . . the practice of each science is by use and not by theory or speculation.

But lecture demonstration in the modern sense of the word probably began towards the latter half of the seventeenth century. The Royal Society was founded in 1660 and, as I mentioned in the Prologue, appointed Robert Hooke as demonstrator; but the first person (according to Desaguliers[11]) to give public demonstrations as part of a course of instruction was John Keill (1671–1721). Records suggest that he began a series of experimental lectures on Newtonian philosophy in 1694. Desaguliers writes

> Dr. John Keill was the first who publicly taught Natural Philosophy by Experiments in a mathematical Manner; for he laid down very simple Propositions, which he prov'd by Experiments. . . .
>
> . . . He began these courses in Oxford about the year 1704 or 1705 and that Way introduc'd the Laws of the Newtonian Philosophy. There were indeed, about the same time, Experiments shown at London by the late Mr. Hauksbee which were electrical, hydrostatical and pneumatical: But as they were only shown and explain'd as so many curious Phaenomena, and not made Use of as a Medium to prove a Series of philosophical Propositions in a mathematical Order, they laid no such Foundation for true Philosophy as Dr. Keill's Experiments; tho' perhaps perform'd more dexterously and with finer Apparatus.

In spite of Desaguliers' strictures I shall have more to say on Hauksbee later.

1.2 DEMONSTRATION IN THE EIGHTEENTH CENTURY

John Theophilus Desaguliers (1683–1744) was born of Huguenot parents who came first to Guernsey and later settled in Islington. He was at one time curator of experiments at the Royal Society as Hooke had been. Again in the preface to his *Experimental Philosophy*[11] he wrote

> When Dr. Keill left the University, I began to teach Experimental Philosophy, after the same Method that he had done, adding the Mechanicks (strictly so call'd, that is, the Explanation of mechanical Organs and the Reason for their Effects).

His book is a marvellous compilation of demonstrations, many of which could be set up and used very effectively in the present day. I shall give two examples: the first forms the basis of a child's toy that is still sold from time to time and is a demonstration that has appeared in many guises since the seventeenth century; the second is a demonstration that is mentioned in §1.5 as the subject of 'fraud' according to Todhunter, but nevertheless is still working very satisfactorily in the Bristol Exploratory[12].

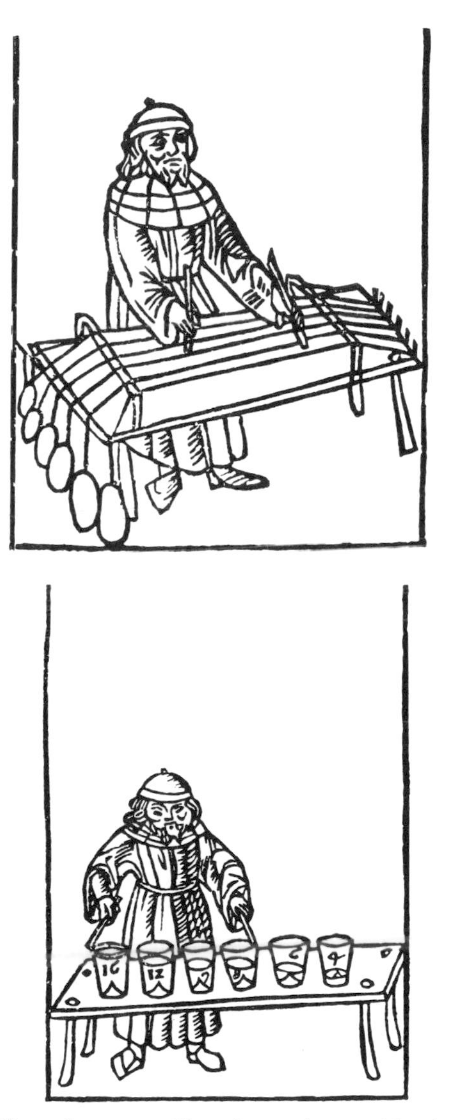

Figure 1.1 Fifteenth century French woodcuts said to be of Pythagoras 'calculant l'echelle des sons'. (BBC Hulton Picture Library.)

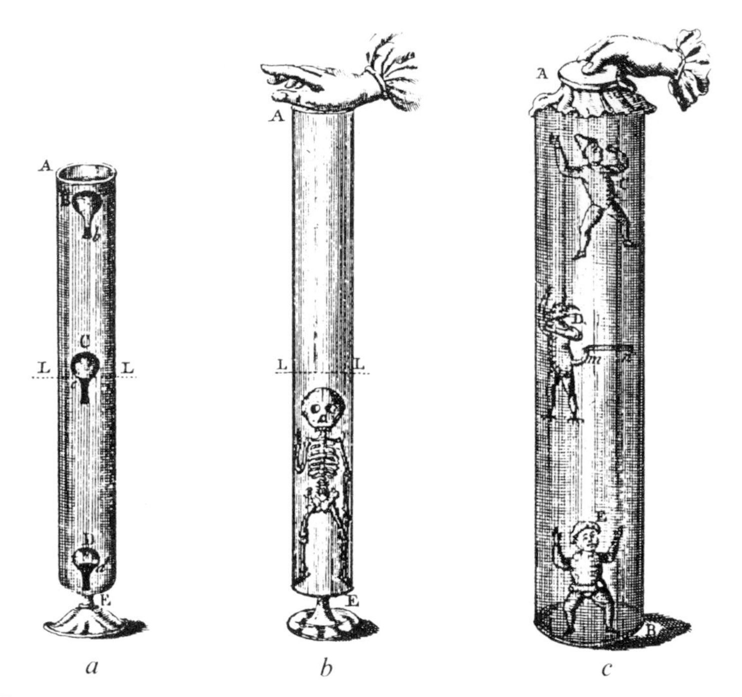

Figure 1.2 The 'diver' experiment as illustrated by Desaguliers. In (*a*), the diver has enough water in the neck to make it only just float; if it is pushed down to positions C or D, more water is forced into the neck and it stays there. In (*b*) and (*c*) the manikin is hollow and has a hole in its base and would normally float comfortably at the surface, but when the pressure is increased through the membrane it sinks, to rise again when the pressure is reduced. (From Desaguliers 1763 *Experimental Philosophy* 3rd edn, figures 12, 18 and 19 of plate 19.)

Demonstration 1.1

A small glass bubble, open at its lower end, is made and weighted so that it only just floats when immersed with its open end down in a tall jar of water. If it is then carefully pushed down to the bottom of the jar it stays there because the air is compressed and water enters the bubble and so increases its total mass. Desaguliers discusses this in detail and then goes on:

> *It is upon this Principle that Glass Images rise and fall in Water; but with the Exception, that they are too light to be made specifically heavier than water, by the Weight of the*

> *Water in the Vessel only; and therefore a Bladder being tied over the Surface of the Water, the Hand which presses upon the Bladder, forces the Water into the Images, and makes them sink: but upon Removal of that Pressure, the Air in the Images forces out the Water by its expansion, and they rise again.*

Figure 1.2 shows the illustration from his book and figure 1.3 shows a Victorian version[13].

Figure 1.3 A Victorian version of the diver. (From *Scientific Mysteries*, 1891, The Chemist & Druggist, 42 Cannon Street, London, p. 73.)

Demonstration 1.2

Figure 1.4 shows the illustration from Desaguliers' book of a wooden machine about ten inches high, two feet long and two inches thick. Between I and F there is a channel in the form of a semi-cycloid, inverted, with an additional horizontal portion about one foot long running from F to G. The channel is made very smooth and is divided into two along its length by a thin, upright brass partition.

> *...The whole instrument may be set upright and horizontal by means of three Skrews such as C,C and the Plummett NM.*

> *Two Brass balls, of half Inch Diameter each, are to move in the two Channels. Fix the two Stops exactly at F and the Brass Balls, though you let them go from different Parts of their respective cycloidal Channels, will strike them at the same Time; which will also be very sensible to anyone who holds a Finger in each channel at F, whilst another person lets the Balls run down from unequal Heights.*

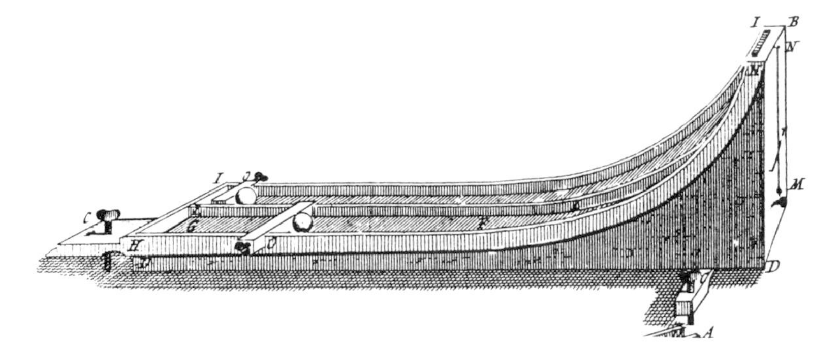

Figure 1.4 Cycloidal track down which two half-inch brass balls can run as illustrated by Desaguliers. (From Desaguliers 1763 *Experimental Philosophy* 3rd edn, plate 27 figure 8.)

If the stops in the channels are moved so that one is four inches beyond F towards G and the other six inches beyond, and the ball in the first channel is released from a point four inches above the horizontal portion and the other from nine inches above the balls should strike the stops simultaneously because

> *Four and Nine are the spaces fallen through, whose Roots two and three express the respective velocities of the Balls, which are shown by one of them running 4 inches in the horizontal Part of its Channel, the other runs 6 inches in the horizontal Channel.*

Figure 1.5 shows another way of using a cycloidal track, in this case a complete inverted cycloid and a straight channel side by side. The straight channel is pivoted about its top end so that it can take up a position such that it crosses the cycloid at its lowest point C, or so that it reaches the horizontal next to the point vertically below the end of the cycloid. In the first configuration, balls released simultaneously from the top of the two channels will separate and that in the cycloidal track will have passed C before the one in the straight track reaches C. In the second configuration the ball in the cycloidal track will move up the second part of the track and return to C, arriving just as the one in the straight track reaches the end.

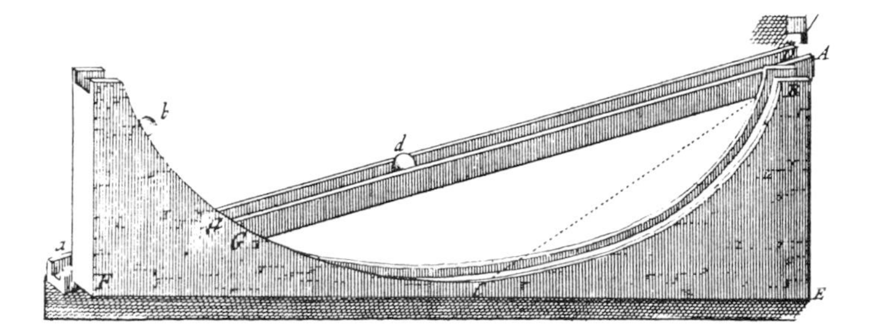

Figure 1.5 Another version of Desaguliers' cycloidal track experiment. (As figure 1.4, plate 27 figure 9.)

Desaguliers commented slightly caustically about Hauksbee's demonstrations not being built into a logical sequence but Hauksbee's own book seems to suggest otherwise. There were, in fact, two Hauksbees, both called Francis and both involved in lecture demonstration. The elder (*c.* 1666–1713) was uncle to the other (1688–1763). The elder wrote a book called *Physico-mechanical Experiments on Various Subjects*[14]. He began as a paid demonstrator at the Royal Society and one of his early demonstrations arose out of his work on a model airpump that he had built and is well worth describing.

Demonstration 1.3

He evacuated a glass reservoir and then allowed mercury to be sucked up a tube into the reservoir and to fall over an inverted glass vessel inside. He described the result[14] *as*

> *. . . a shower of Fire descending all round the Sides of the Glasses*

This was round about 1705, and he went on to explore this phenomenon and to try all kinds of variations in order to discover the source of the luminosity. In particular he set up a small glass globe that could be rotated rapidly against a piece of woollen cloth inside the evacuated receiver. This

> *...quickly produced a beautiful Phaenomenon, viz. a fine purple light, and vivid to that degree, that all included Apparatus was easily and distinctly discernible by the help of it.*

Among many other series of demonstrations and experiments he also explored surface tension and was probably the first to perform what has become a classic, though very simple, demonstration of capillary rise.

Demonstration 1.4

Two rectangular glass plates are placed together so that an edge of each is in contact with the other and the common edge is vertical. The opposite edges are separated slightly to form a wedge-shaped air space between them. The lower edges of the pair are then immersed just below the surface of coloured water. The liquid, of course, rises more in the thinner parts of the wedge and the upper surface forms a hyperbola which is asymptotic at one end to the liquid surface in the vessel and at the other end to the vertical contact line of the two plates (figure 1.6). Among other explorations Hauksbee went on to show that this effect was totally independent of the thickness of the plates, even if they were made up to ten times thicker.

In Germany, Johannes Andreas Schmidt taught physics at the university in Helmstedt at the begininning of the eighteenth century and he published a textbook with the title *Collegii Experimentalis Physico-mathematici Demonstrationes* and which contained many pictures and diagrams of apparatus. Professor R W Pohl[15] says that most of what Schmidt discussed in his chapters on mechanics, acoustics and thermodynamics, such as lifting devices, megaphones and diving bells, would be covered in an engineering course today. The optics is unclear and electricity is not mentioned. There are, however, some curiosities included, as for example the behaviour of a drop of molten glass dropped into water, which shatters when the 'tail' is subsequently broken. A brief description of this effect that appeared in 1891[16] is worth quoting.

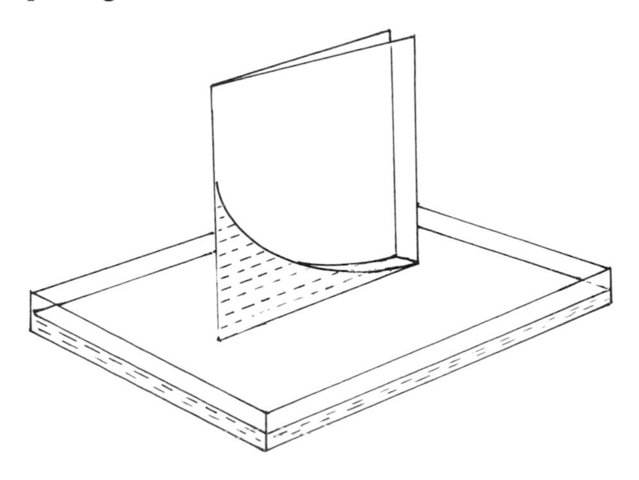

Figure 1.6 Hyperbolic surface of a liquid in the wedge-shaped space between two glass plates.

Demonstration 1.5

Prince Rupert's drops is the name given to tears of glass which have been suddenly cooled by dropping into water. On nipping off the end they fly into impalpable powder, and if held firmly in the hand produce a most curious sensation. Place one of these drops into a tumbler of water and nip off the smaller end—the explosion that follows is so energetic as to even break the glass vessel.

(N.B. The anonymous author of this book, which cost one shilling, does not suggest any safety precautions!)

Georg Christoph Lichtenberg (1742–99) began to teach experimental physics at Göttingen in 1781, and the first university chair of experimental physics in Germany was specially created for him. He is probably best known for his demonstration of the figures that arise when an electrical discharge is created near or in contact with a plate of insulating material dusted with powdered insulator. This has had enormous consequences . . . it is the fundamental principle of the modern photocopier and has also been used in the study of electrical discharges. These so called 'Lichtenberg figures' were also the inspiration for Chladni's sand figures which are described in §3.6.3, Demonstration 3.6. Lichtenberg believed in large apparatus and said that

> Repeating an experiment with larger apparatus is tantamount to looking at the phenomenon through a microscope.

Merrill and von Hippel[17] record that

> faithful to his maxim of exploring nature with instruments of unusual dimensions, [he] baked a tremendous resin cake for electrostatic experiments

By the middle of the eighteenth century lecture demonstration was becoming widespread. In his book *Physics Teaching in England from Early Times up to 1850* G D Bishop[18] quotes two interesting items. The following advertisement appeared in the *Daily Advertiser* in 1746:

> At the Swan Tavern in Exchange Alley, Every Hour from Eleven in the Morning till Five in the Afternoon, will be exhibited A Regular series of about forty Experiments with concise Observations by way of Lecture; illustrating the known principles and demonstrating the Activity and Force of Electricity; in Attraction and Repulsion; producing Light and actual Fire from various Substances especially Human Bodies . . . By the method of this Course the whole Affair of Electricity is pursued from its simplest to its most capital Phenomena.

And in 1792 in *An Address to the Ingenious Youth of Great Britain*, J C Ryland wrote that lack of apparatus is no justification for teaching physics only out of books:

> . . . a fire shovel, tongs and poker, will show the foundation of the mechanic powers; especially the nature of levers. A

spinning wheel will clearly show the power of wheel and axle . . . marbles will teach the schoolboy the nature of percussion and the laws of motion . . . A schoolboy's jews harp will serve to teach us those tremulous motions which are the cause of sounds; and a glass prism, a looking glass and an ox's eye from the butcher's will be a happy foundation of optics.

1.3 DEMONSTRATION IN THE EARLIER PART OF THE NINETEENTH CENTURY

It would, of course, be quite impossible to describe all the remarkable demonstrations that were developed in the nineteenth century. It was a tremendously rich time and many of the great demonstrations first introduced during that period are described in later parts of this book in order to illustrate a particular aspect of the technique of lecture demonstration. My intention here and in the next section is to pick out some that both illustrate the attitudes of the demonstrators concerned and also have become, in some form, part of the modern repertoire.

I shall start with Thomas Young (1773–1829). Unfortunately he was not a very competent lecturer, but he fully justifies a place here because of two demonstrations that are still regularly performed in various different ways that evolved out of his experiments. The first is still given his name—Young's fringes—and the second, based on an experiment done by Young in his studies of colour vision, is often known as Young's colour patch experiment.

Demonstration 1.6

Young was an ardent supporter of the wave theory of light, though he argued, by analogy with sound, that it was probably longitudinal. His famous experiment, often known as the 'double-slit interference' experiment was, in fact, first performed with pin-holes. He allowed a beam of sunlight to fall on a single pin-hole to give the diffraction effect, first observed by Grimaldi (1618–63), and then to place two pin-holes very close together so that they were illuminated by the central patch of the diffraction pattern of the first hole. The result was that where the light from the two pin-holes overlapped, coloured interference bands occurred. Nowadays, we usually perform the experiment with a single slit followed by a double slit, which gives fringes that are more easily visible. For large-scale demonstration a laser source works well and, because laser light is already spatially coherent, the first slit is unnecessary.

Young demonstrated the nature of the interference phenomenon using a ripple tank, and the original one that is believed to have belonged to Young is still in existence at the Royal Institution. He may not have been a very competent lecturer himself, but his double-slit experiment must stand as one of the all-time great contributions to the art of lecture demonstration.

Demonstration 1.7

Three projectors are needed for this experiment: each carries a slide giving a circular patch of light of one of the primary colours, red, blue and green. The projectors are adjusted so that the patches overlap (see figure 1.7) giving white in the central triangular patch and the three subtractive primaries (minus red = cyan, minus green = magneta, and minus blue = yellow) in the three areas where just two primaries overlap.

In §3.6.4, Demonstration 3.7, various modifications of this experiment are described.

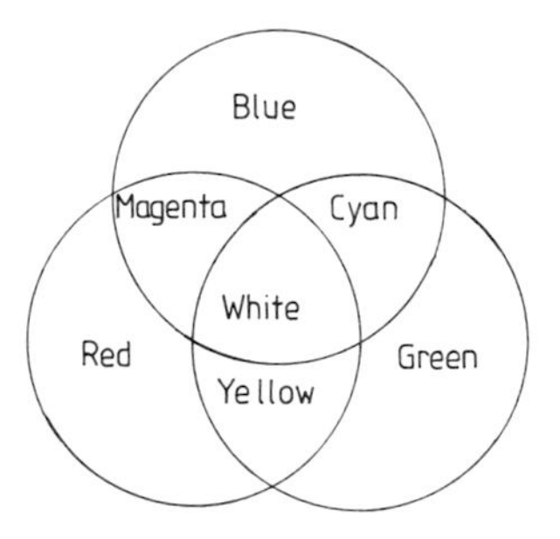

Figure 1.7 Appearance of the screen during Young's colour patch experiment.

Sir Humphry Davy (1778–1829) is remembered by the public chiefly for his invention of the miners' safety lamp. But he made enormous contributions to science in many other ways, including virtually doubling the number of chemical elements that could be isolated, and, of course, in 1812, what he himself described as his greatest discovery—Michael Faraday. I shall describe here two of Davy's most spectacular experiments, which are still regularly performed.

Demonstration 1.8

Figure 1.8 shows the apparatus as it is illustrated in John Tyndall's Heat; a Mode of Motion[19]. *Davy placed a basket of glowing charcoal at the focus of the upper mirror and this caused the explosion of a small quantity of gunpowder placed at the focus of the lower mirror when it was moved into the right position to produce a focused 'image' of the charcoal formed by the heat radiation. Tyndall used to perform a variation of this experiment which demonstrated the focusing of light and of radiant heat separately. He placed a transparent collodion balloon filled with a mixture of hydrogen and chlorine at the focus of the upper mirror and, when a carbon arc was struck at the focus of the lower mirror, the balloon exploded, but the remains of the balloon floated down unburnt, showing that the light had caused the gases to combine. The balloon was then blackened and filled with a mixture of oxygen and hydrogen and the experiment repeated; this time the explosion occurred as before but the balloon was completely burnt. The mirrors that are still used for this experiment at the Royal Institution are believed to be the originals; it is interesting to note Tyndall's footnote to his description:*

> *In an excellent magazine, published in Glasgow when I was a youth, I first read an account of Davy's experiments; and to the present hour I remember the yearning which took possession of me to be, like him, a natural philosopher. I had little notion at the time that I should ever work with the very instruments the description of which had so fired my young enthusiasm.*

The second of Davy's experiments that I want to describe seems to me to combine all the virtues of the ideal demonstration experiment. It is simple, has strong visual appeal and makes its scientific point very clearly without the need for complicated explanations.

Figure 1.8 An experiment on the focusing of radiant heat by parabolic mirrors as performed by Davy. (From Tyndall J 1898 *Heat; a Mode of Motion* 11th edn (London: Longman) p. 290, figure 92.)

Demonstration 1.9

This experiment is designed to illustrate the fact that the electrical resistance of a metal decreases if the temperature is reduced. A length of wire is stretched between two terminals about a metre apart and a current is passed through it. The current is adjusted so that the wire gets hot, but does not quite begin to glow visibly. (Davy had acquired a large battery of zinc–copper cells, paid for by special subscription of the members of the Royal Institution.) A piece of ice is then rubbed backwards and forwards along a central portion of the wire about 30 cm long. As the wire cools the two portions on either side of the cooled portion begin to glow red hot. The effect is quite dramatic in a darkened room. Clearly the resistance of the cooled portion has dropped very considerably, thus increasing the current flow.

My last exemplar of the demonstration lecturer in the first half of the nineteenth century is Michael Faraday (1791–1867). He followed Davy as Superintendent of the House at the Royal Institution (see §1.7) and left some remarkably good advice to lecturers that is quoted in many places in this book[1]. The choice of which of his demonstrations to describe is very difficult and I have selected four which I hope will give a flavour of the kinds of experiments he did. I may well be criticised because I have chosen experiments in physics. But, as I said earlier, I am a physicist and would not feel competent to write about demonstrations in another field. But I must make the point that a high proportion of Faraday's demonstration experiments were in chemistry; indeed his own first series of Christmas Lectures was entitled 'The *Chemical* history of the Candle'. The first experiment I shall describe could, I suppose, be described as either chemical or physical, but whichever category it comes in, it was one of his favourites, and is one that is still done to good effect

Demonstration 1.10

The experiment is to show that the gases given off by the electrolysis of acidulated water are hydrogen and oxygen in the proportions in which they make up water and hence constitute an explosive mixture. A cell closed by a cork and containing

acidulated water and two electrodes has a current passed through it and the mixed gases emerge from a tube passing through the cork. Soap solution (nowadays washing-up liquid is easier to use) is placed in the palm of the lecturer's hand and a small bubble blown with the gases emerging from the tube. The bubble explodes on the lecturer's hand when a lighted match is applied. This experiment still causes considerable amusement and interest to modern audiences. It is perhaps worth noting that I had some trouble a year or two ago in getting the experiment to work. The problem was that my cell did not produce much gas and did produce a thick white precipitate that soon coated the electrodes. The problem turned out to be the presence of some constituent of the water and was cured by using de-ionised water acidulated with a little sulphuric acid.

The next demonstration arises out of Faraday's work on low temperatures. He was able to liquefy all the known gases, except the so-called 'permanent' gases and to solidify carbon dioxide.

Demonstration 1.11

A platinum crucible is raised to red heat and then some solid carbon dioxide and a little liquid ether (to increase the thermal contact) are placed in the crucible; a metal spoon containing mercury is then lowered into the mixture. Although the crucible remains red hot, the layer of carbon dioxide gas acts as an insulator between the crucible and the freezing mixture and it remains cold long enough to freeze the mercury.

There are obviously hazards in performing this experiment: indeed John Tyndall (who was always scrupulously honest in giving accounts of his demonstrations, and always ready to turn disaster to advantage) performed this demonstration and wrote[20]

> *. . . I dip a conical spoon, containing the liquid metal, into the red-hot crucible, and surround it . . .with carbonic acid and ether. The ether vapour has taken fire, which was not intended. The experiment ought to be so made, that the carbonic acid*

gas—the choke damp of mines—shall preserve the ether from ignition. The mercury, however, freezes, the presence of the flame adding to the impressiveness of the result, as the intensely cold solid is lifted through the fire.

The third example is of a demonstration that illustrates a principle that is made use of in just about every modern electronic device in existence—the Faraday cage. The description is of the modern version of the demonstration as performed, for example, by Sir George Porter.

Demonstration 1.12

A large cage of wire netting, big enough for a person to sit inside, and fitted with a door to permit entry, is mounted on an insulating base. A volunteer sits inside the cage, the door is closed and the cage is charged to a high potential by means of a large Whimshurst machine. Paper streamers attached to the outside of the cage stand out perpendicularly to the walls, showing the direction of the electric field outside the cage, but a streamer on a stick held by the volunteer and the volunteer's hair are unaffected because, as Faraday pointed out, the cage forms a conducting surface and so there can be no differences of potential between its various parts, and hence no field can exist inside. If the volunteer pokes the stick through the netting, the streamers immediately stand up.

Finally I shall describe another of Faraday's experiments the consequences of which have recently assumed enormous importance in the field of communications and the new science of optoelectronics.

Demonstration 1.13

The original apparatus used by Faraday still exists in the Royal Institution (figure 1.9(a)) and is illustrated diagrammatically in figure 1.9(b) which is taken from Tyndall's Light[21]. *N is the projection lens of the light source, H a Nicol prism, Q a bi-quartz plate (one half is left handed and the other right handed), L a lens, M the electromagnet with a cylinder of Faraday's 'heavy glass' mounted in its perforated poles and P the second Nicol prism. With the current switched off the polariser is turned until the two halves of the field appear tinged exactly the same colour (which is described by Tyndall as 'puce'). When the current is turned on, the plane of polarisation is rotated and, as a result, one half of the field becomes slightly pink and the other half greenish. When the current is reversed the two colours are interchanged.*

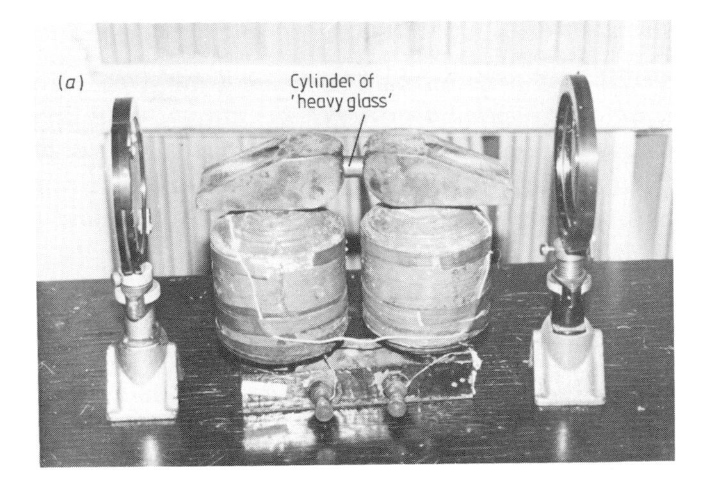

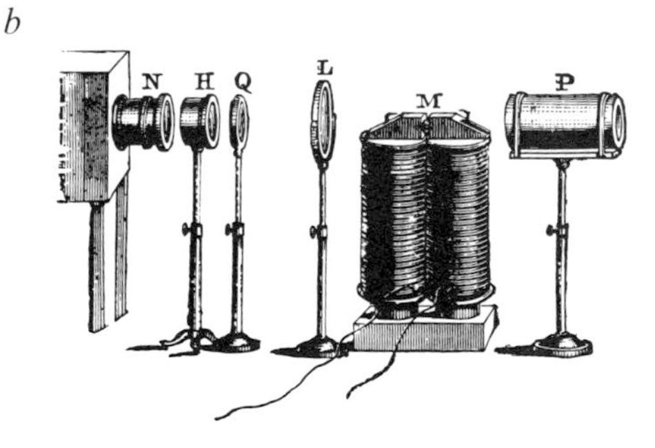

Figure 1.9 (*a*) Faraday's magnet with pierced pole pieces and cylinder of heavy glass for showing the rotation of the plane of polarisation of light by a magnetic field as it exists today at the Royal Institution. (From Taylor C A 1986 *Proc. R. Inst.* **58** 90, photo by R S Watkins.) (*b*) Diagram of (*a*) from Tyndall's '*Six Lectures on Light*'.

1.4 DEMONSTRATION IN THE LATTER PART OF THE NINETEENTH CENTURY

John Tyndall (1820–93) clearly fits in this section because he started his lecturing career in 1853; but in another sense he forms a bridge between the two parts of the nineteenth century. He was a great admirer of Faraday and others from the earlier period and many of his demonstrations are improvements and modifications of ideas stemming from the earlier years.

I have to confess that Tyndall is a particular favourite of mine and perhaps apologise for the frequency of references to him throughout the book. I think the justification is that he took up and developed ideas from many different sources and studying his lecture demonstrations alone gives a very good picture of the state of the art at that time.

He also left very detailed records of his demonstrations. For example there is a bound volume of manuscript notes[22] in the archives at the Royal Institution that formed the basis for a discourse that I gave on Tyndall's demonstrations on sound and which gives a splendid insight into Tyndall's character. He seemed to thrive on controversy, but quite obviously loved winning an argument. In 1854 he engaged in an argument with Forbes over the 'Trevelyan Rocker'. The rocker itself is of scarcely more than academic interest. It is a bar of triangular section and, depending on the dimensions, when it is heated and then placed on its supports it will rock either visibly or so fast as to emit a note which can be changed by pressing on the bar. Forbes had developed three 'laws' which he claimed governed the behaviour and Tyndall proceeded to demolish them experimentally, one by one. The relevant pages of the manuscript are shown in figure 1.10.

I shall describe four of Tyndall's demonstrations here, though others are described elsewhere in the book. The first concerns radiant heat, the second his famous experiments on singing and sensitive flames, the third the conduction of sound through solids and the fourth his demonstration of the presence of acoustic strain using polarised light.

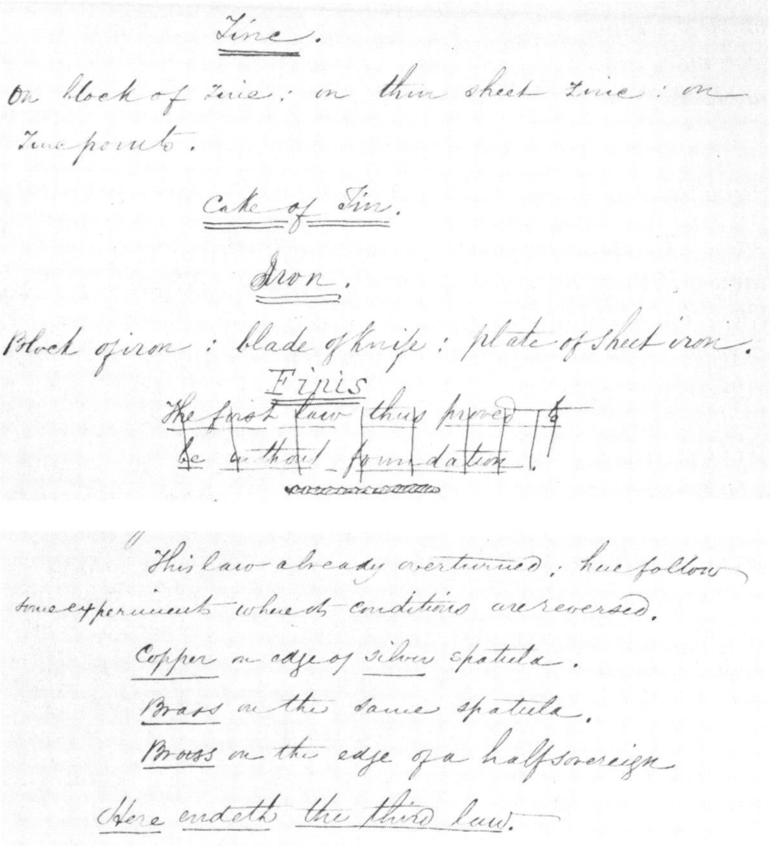

Zinc.

On block of zinc: on thin sheet zinc: on zinc points.

Cake of Tin.

Iron.

Block of iron: blade of knife: plate of sheet iron.

Finis

~~The first law thus proved to be without foundation!!~~

This law already overturned; here follow some experiments where the conditions are reversed.

Copper on edge of silver spatula.

Brass on the same spatula.

Brass on the edge of a half sovereign

Here endeth the third law.

Figure 1.10 From Tyndall's manuscript notes for 27 January 1854 relating to Forbes' three laws governing the behaviour of Trevelyan's rockers. (Royal Institution archives.)

Demonstration 1.14

A round-bottomed flask is filled with water and used to focus a spot of light from an arc on to a piece of black card. Because the water absorbs the heat radiation there is no effect on the card.

An identical flask filled with carbon tetrachloride in which iodine has been dissolved to make it completely opaque to visible light is then substituted for the water-filled flask. The black card begins to burn as a result of the focused heat radiation for which the iodine solution is transparent. The effect can be made even more dramatic if a little gunpowder or other very inflammable material is attached to the card.

Demonstration 1.15

One of Tyndall's most famous experiments involves the use of flames. It must be remembered that Tyndall worked at a time when the means of detecting sounds were somewhat restricted; there were no electronic amplifiers and, certainly in his early days, no microphones. He found that a flame produced from a small-diameter gas jet could act as a detector under certain circumstances; if the gas pressure is increased there would come a point when the streamlined flow, giving rise to a long thin flame, would break down and the flame would shorten and spread out. But if the pressure is adjusted until the flame is just on the point of shortening it becomes very sensitive to any external disturbance. Any high-frequency sound such as the rattle of a bunch of keys, or the sibilants in a word makes the flame 'duck'. Indeed in Tyndall's notes[22] *he gives some suggestions for sentences that might be suitable including a passage from Spenser concluding with the line 'A silver sound, which heavenly music seemed to make'.*

In attempting to reproduce this experiment I use a piece of glass tubing drawn down to about 1 mm diameter as the jet and have found that the flame can be made stable over a longer length and hence more sensitive by packing the wide part of the glass tube below the jet, over a length of perhaps 100 mm or so, loosely with glass wool. This seems to 'iron out' some of the departures from streamlined flow resulting from irregularities in the gas flow. I have also found it difficult to operate with natural gas and usually use 'simulated town gas' which can be obtained from the usual gas cylinder supplier.

Tyndall also used smoke jets and liquid jets in a similar way. The liquid jet experiment can be very dramatic. A thin jet of water from a constant head supply is made to flow from a thin glass jet comparable with that used for the flame experiment. The jet is tilted so that it describes a parabolic path and falls on to a thin, tightly stretched, horizontal rubber membrane about 100 mm in diameter placed about a metre from the orifice. If a watch is now placed in contact with the glass jet near to the orifice, the ticks disturb the flow and the sound of the jet striking the membrane is modulated so that the ticks of the watch can be heard by a large audience.

The other adaptation of the flame experiment that works extremely well is the so-called 'singing flame' experiment. Figure 1.11 shows the arrangement. A small gas flame about 8–10 mm long is created at the end of a tapered metal tube about 150–200 mm long and a glass tube about 20 mm in diameter and about 300 mm long is mounted vertically with the gas jet inside the lower end. If the tube is carefully raised and lowered a point can be found at which the tube emits a loud note and the flame becomes elongated. Again I have found it better to use simulated town gas and the length of the metal tube turns out to be critical in relation to the length of the glass tube. A cardboard slider placed over the end of the tube and raised and lowered (as in the diagram) alters the pitch of the note. An elegant variation is to adjust the height of the tube so that the flame is just not *singing; if you then sing precisely the right note to create resonance the tube begins to sing in sympathy, but, having started, it continues until the tube is raised again. It is possible to persuade the tube to sing in sympathy from a considerable distance; but of course if the note sung by the demonstrator is not the right one the flame does not respond. Tyndall's own account*[23] *makes amusing reading (though I must confess that I have never succeeded in getting the flame to respond from more than a few feet away).*

> *Give me your permission to address that flame. If I be skilful enough to pitch my voice to a certain note, the flame will respond by suddenly starting into melodious song, and it will continue*

singing as long as the gas continues to burn. . .I emit a sound, which you will pardon if it be not musical. The flame does not respond, it has not been spoken to in the proper language. But a note of somewhat higher pitch causes the flame to stretch, and every individual in this large audience now hears its song. I stop the sound, and stand at a greater distance from the flame: now that the proper pitch has been ascertained, the experiment is sure to succeed and from a distance of 20 or 30 ft the flame is caused to sing. . . .

In order to appreciate the impact on an audience of my third example of a Tyndall demonstration it is necessary to remember that in the 1850s when this was first performed the loudspeaker and the gramophone and similar devices had not yet been invented and the only sources of musical sounds were musical instruments themselves. The idea of the experiment was apparently first suggested by Sir Charles Wheatstone[24].

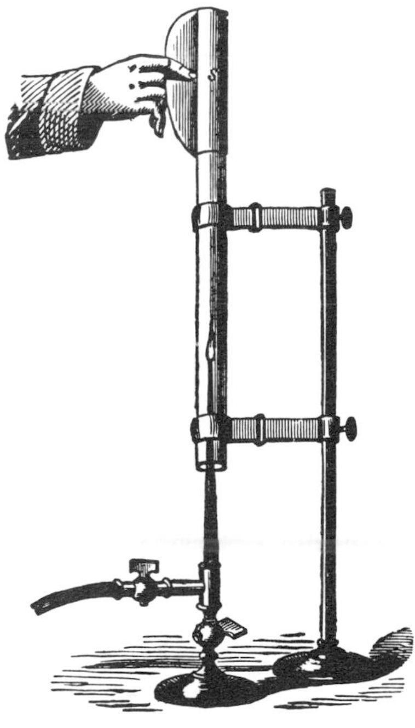

Figure 1.11 Tyndall's singing flame apparatus. (From Tyndall J 1875 *Sound* 3rd edn (London: Longman) p. 248, figure 115.)

Demonstration 1.16

A hole was made in the floor of the lecture theatre and a tin tube 2.5 inches in diameter was passed down through two floors to the basement. Along the axis of this tube a deal rod was supported by rubber so that it did not touch the tube. The bottom end rested on the sound board of a grand piano in the basement with a pianist performing. The upper end emerged in front of the lecture bench and only an inch or so above the floor. When a violin, cello, guitar or harp was placed on the end of the rod, to the audience's amazement the sound of the piano poured forth from it. Tyndall's own description of the result of the experiment is a classic of Victorian florid prose and is well worth reproducing:

> *What a curious transference of action is here presented to the mind! At the command of the musician's will, the fingers strike the keys; the hammers strike the strings, by which the rude mechanical shock is converted into tremors. The vibrations are communicated to the sound-board of the piano. Upon that board rests the end of a deal rod, thinned off to a sharp edge to make it fit more easily between the wires. Through the edge, and afterward along the rod, are poured with unfailing precision the entangled pulsations produced by the shocks of those ten agile fingers. To the sound-board of the harp before you, the rod faithfully delivers up the vibrations of which it is the vehicle. This second sound-board transfers the motion to the air, carving it and chasing it into forms so transcendently complicated that confusion alone could be anticipated from the shock and jostle of the sonorous waves. But the marvellous human ear accepts every feature of the motion, and all the strife and struggle and confusion melt finally into music upon the brain.*

An extremely complicated version of this experiment described as 'Wheatstone's telephonic concert' was performed at the Polytechnic with several different instruments in the basement. Figure 1.12[25] shows an elegant parlour variation; it is interesting to note that a modern version, in which sound was transmitted (along the original hole, made for Tyndall!) by means of a laser beam was performed during the 1987 Royal Institution Christmas Lectures.

Figure 1.12 A drawing-room version of Wheatstone's 'telephonic concert'. (From Wylde J *c.* 1860 *Magic of Science* (Griffin) p. 238.)

The final example of a Tyndall demonstration involves the use of polarised light to display internal strains in glass.

Demonstration 1.17

This demonstration requires a strip of glass about two metres long, 50 mm wide and 6 mm thick. (To avoid any possibility of accidental cuts I have the edges ground when I repeat this experiment.) Tyndall's arrangement, using Nicol prisms, is shown in figure 1.13, but it can be done equally well with two pieces of Polaroid instead. The important point is to work with the Nicols, or Polaroids, crossed so that no light is transmitted, and then to arrange that the length of the glass strip makes an angle of 45° with the planes of the polariser and analyser. The strip is clamped firmly at its mid point and then stroked with a wet cloth. The longitudinal vibrations produce a high-pitched note and at the same time the strain makes the glass become anisotropic and light appears on the screen. The secret of good strong vibrations is to have the cloth and the glass very wet. As with other experiments using vibrations of wet glass, the water in some regions is not as effective as in others because of differences in the mineral content. To avoid this problem when visiting a new locality I take with me a bottle of proprietory mineral water which I know works satisfactorily.

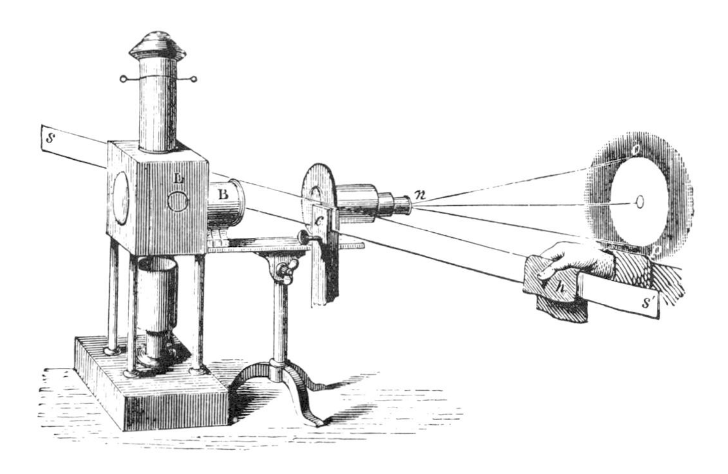

Figure 1.13 Tyndall's apparatus for demonstrating acoustic strain using polarised light. (From Tyndall J 1882 *Light* 3rd edn (London: Longman) p. 136, figure 39.)

Sir James Dewar (1842–1923) performed some spectacular demonstrations during his period at the Royal Institution (beginning in 1877). One of these is commemorated in an oil painting, exhibited outside the lecture theatre, that I find quite disturbing (figure 1.14). It shows Dewar demonstrating the properties of liquid hydrogen in 1904 in front of a distinguished audience, and one cannot help wondering about safety precautions! One of his experiments that would have delighted Faraday (who knew that gaseous oxygen was magnetic) was to pick up liquid oxygen with an electromagnet.

Figure 1.14 Dewar lecturing on liquid hydrogen at the Royal Institution, painted by T Brooks in 1904. (Courtesy of the Director of the Royal Institution.)

Before leaving this period there are two other scientists that I should like to mention, both of whom contributed greatly to the repertoire of demonstrations in sound; one is Hermann von Helmholtz (1821–94) who, together with Karl Rudolph König (1832–1901), designed some of the most superb apparatus, and the other is Dayton Clarence Miller (1866–1941) whose Phonodeik enabled so much progress to be made in understanding the nature of musical sound waves (strictly this did not appear until about 1908, but it really belongs to this earlier period).

Helmholtz worked in a staggering variety of fields: conservation of energy, electrical inductance, anatomy and physiology, psychoacoustics and many more. His great work *On the Sensations of Tone*[26] is still required reading for anyone beginning the study of psychoacoustics. He held posts at many universities in Germany, in departments of anatomy, physiology and in physics, finally being invited to the Chair of Physics in the University of Berlin in 1870. I shall just describe two of the pieces of equipment that were designed by him for research purposes, but subsequently produced by König for sale, largely for demonstration purposes.

Figure 1.15 The double siren as manufactured by König to Helmholtz's design (University College, Cardiff). Rotation of the white handle on the upper siren alters the angular position of the fixed plate and hence the phase of the pulses produced relative to those produced by the lower siren. (Photo by R S Watkins.)

Demonstration 1.18

The first is his double siren; figure 1.15 shows the instrument. The two sirens are mounted on the same shaft: the upper has four rings with 9,12,15 and 16 holes respectively; the lower has four rings with 8,10,12 and 18 holes respectively. If the tone given by the eight-hole row is taken as C then the upper disc gives D,G,B and C′, and the lower, C,E,G,D′. Any ring can be opened or closed by pulling or pushing the tangential rods and so a whole series of pairs of notes giving precise musical intervals ranging from unison, semitone, etc, up to fifths and octaves can be obtained. This was the first instrument that gave absoloutely precise ratios of frequency and remained the only really precise one until the advent of digital synthesis. An added refinement was that, by means of the handle and cog wheel at the top, the chamber on the top of which the rings of fixed holes are drilled could be rotated, hence altering the phase relationship between the notes produced by the top disc and those by the bottom. Helmholtz used this instrument in his investigations of the origins of consonance and dissonance and the nature of combination tones.

Figure 1.16 Set of Helmholtz resonators (Courtesy the Royal Institution.)

The tones produced by the siren are obviously not pure; they are approximations to square waves, but their fundamentals are precise in frequency. But, in order to do experiments on pure tones, some kind of filter is needed to enable the experimenter to listen to a given sinusoidal component at will. Helmholtz designed his famous resonators to do just this. Each resonator was precisely tuned so that one component of each compound note could be heard when the tapered nozzle was placed in the ear. Figure 1.16 shows a set of these resonators as manufactured by König. In the early part of the present century there were few university physics laboratories which did not possess such a set.

The American physicist D C Miller developed an instrument that he called the Phonodeik for displaying waveforms produced by musical instruments and the human voice.

Demonstration 1.19

The essence of the instrument is shown in figure 1.17. This was the first really successful method of observing waveforms and was widely used for demonstration in schools and universities. Indeed I remember that my school physics laboratory posessed one and I also remember Dr Alexander Wood using one, with a rotating mirror drum to spread the waveform of Rachmaninov's G minor Prelude all round the walls of the old Cavendish lecture theatre in Cambridge in 1941; it was a most impressive demonstration. It is only of historical interest now because the display can be done so much more easily and precisely by electronic means, but I felt it important to mention as a milestone in the development of lecture demonstrations.

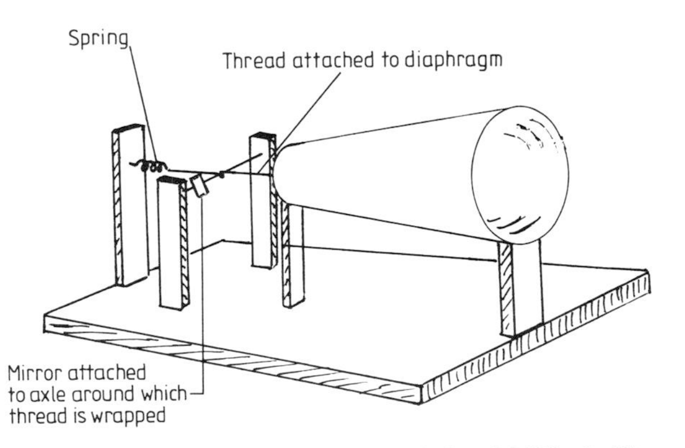

Figure 1.17 Diagram showing the principle of Miller's Phonodeik.

1.5 CONTROVERSY ABOUT THE VALUE OF DEMONSTRATIONS

There were plenty of opponents in the latter half of the nineteenth century who felt that demonstration was at best a waste of time and at worst could be harmful to the educational process. I quoted one extract from Galloway's writing in this vein in the Prologue but he also had a number of other pungent things to say that must be mentioned in order to give a fair picture. He asserts that one cannot lecture on a scientific topic without using the technical language of that topic[27]:

> Scientific men cannot bridge over the gulf for the non-scientific by stripping a lecture on any science of the language of that science; for when that is done, it ceases to be science, and can only be, whether by experiments or other aids, an amusement for the hearers. Hence one of the reasons why Popular Scientific Lectures never can be a means of instructing those who neither know the language of science, nor yet have any ideas wherewith to connect and blend what they hear, so that it becomes *knowledge* and of a progressive character.

He goes on with his castigation of popular lecturers by suggesting that the

> . . . so-called science poured forth by the lecturer is not always of the soundest description; and the experiments shown are said to be sometimes deceptions

I hope such an accusation cannot be made about today's demonstrations. Isaac Todhunter was another opponent of the use of demonstration. In fact Galloway goes on to quote him[28]:

> So great is the difficulty in persuading experiments intended to be visible to a large company to conduct themselves properly that curious charges of unfairness are in circulation, which are more or less authenticated. Thus it is said one lecturer was accustomed to show by experiment that a body would fall down a tube in the shape of a cycloid faster than down a tube of another shape, corresponding to the same vertical height; but in order to assist Nature he was

> wont to *grease* the ball surreptitiously which travelled on the cycloid. Again another lecturer was accustomed to illustrate a mechanical principle known by the name of virtual velocity; a certain weight ought to remain immovable, though not absoloutely fixed; in this case a *nail* applied to the weight, unknown to the spectators, prevented any casualty in the experiment.

It is interesting to note in passing that the cycloid experiment (see also Demonstration 1.2) now forms one of the exhibits in the Exploratory, the interactive science centre in Bristol (see §1.9), where two small cars travel down a cycloidal and a linear track side by side and the experiment is performed by children themselves and works perfectly well without any resort to grease!

But, in the same volume[28], Todhunter objects to experimental demonstration on other grounds:

> We assert that if the resistance of the air be withdrawn a sovereign and a feather will fall through equal spaces in equal times. Very great credit is due to the person who first imagined the well known experiment to illustrate this; but it is not obvious what is the special benefit now gained by seeing a lecturer repeat the process. It may be said that a boy takes more interest in the matter by seeing for himself, or by performing for himself, that is by working the handle of the air-pump: this we admit, while we continue to doubt the educational value of the transaction. The boy would probably take much more interest in foot-ball than in Latin grammar; but the measure of his interest is not identical with that of the importance of the subjects. It may be said that the fact makes a stronger impression on the boy through the medium of his sight, that he believes it the more confidently. I say that this ought not to be the case. If he does not believe the statements of his tutor—probably a clergyman of mature knowledge, recognized ability, and blameless character—his suspicion is irrational, and manifests a want of the power of appreciating evidence, a want fatal to his success in that branch of science which he is supposed to be cultivating.

I beg to differ! But perhaps I might remind readers that Todhunter was a mathematician!

Another of the controversies about demonstration arose

out of H E Armstrong's advocacy of the heuristic approach to science teaching, sometimes known as the 'self-discovery' method (see for example W H Brock's *H E Armstrong and the Teaching of Science, 1880–1930*[29]). But, as in so many other cases, some of the argument arose out of misunderstanding. Armstrong was accused by some people of suggesting that children should rediscover for themselves all that they could be expected to know of science. This is clearly impossible and impractical. In fact Armstrong wanted children to see for themselves how discoveries were actually made, which of course is a different matter. But the supporters of the heuristic method tended to imply that demonstrations were of little value compared with laboratory experimentation by the pupils themselves. Naturally I am wholeheartedly in agreement with the idea that pupils and students should do experimental work themselves; but I hold to the view that practical work in the laboratory and lecture demonstration are not mutually exclusive and really fulfil complementary functions.

Of course the problems of teaching science become particularly acute when the spectre of examination appears. Although I hope that teachers and lecturers working towards examinations of one kind or another will find help in this book, I admit that I have tended to address my remarks rather more to those engaged in the process of increasing interest and understanding of science in a more general way. I remember once hearing an Oxford arts don complaining about the difficulty of examining a real understanding of a poem rather than mere ability to give a critical analysis. She said that she wished it were possible to set a question such as 'Enjoy the following poem', I suppose that would be the equivalent of Armstrong's hope that pupils could be put in the position of the original discoverer of a scientific principle or effect. I find it extraordinary that in so many fields people seem to be unable to accept elements of truth from a number of sources: so often there is complete polarisation and two useful and complementary ideas are seen as diagonally opposed. E W Jenkins[30] points out that reports published by the Science Masters' Association after the Second World War advocated the use of demonstration experiments in teaching science, but in 1950 it was necessary to produce another report pointing out that this advocacy did not imply that demonstration experiments could *replace* laboratory work done by pupils for themselves.

1.6 DEMONSTRATION IN THE TWENTIETH CENTURY

The twentieth century has seen many contradictory developments in lecture demonstration. On the one hand some of the really great exponents of the art (for example Sir Lawrence Bragg) have worked in this period, and some of the most exciting developments in technology (such as closed-circuit television, laser light sources, oscilloscopes, etc) have become available to enhance demonstrations. But, on the other hand, shortage of time, and of technical assistance, have reduced the opportunities for demonstration in the course of ordinary school or university classes. Nowadays the art tends to be fostered only for special occasions, such as Christmas lectures and popular lectures to encourage wider interest in science.

There are many great exponents of the art in the twentieth century and, whichever I choose, I shall be criticised for omissions. I propose therefore to pick out only about half a dozen outstanding demonstrations which make their appeal chiefly because of the simplicity of the idea behind them and because, in spite of that, they make powerful teaching tools and can be used either in popular lectures or in school or university courses. More twentieth-centruy demonstrations are described in later parts of the book to illustrate various aspects of the theme.

The first was devised by Sir Lawrence Bragg.

Demonstration 1.20

In my view the bubble model of crystal structure is one of the most elegant and beautiful demonstrations ever conceived, and yet it is extremely simple and not only works well for demonstration but has contributed ideas in research. A small jet is fed from a constant-pressure air supply and its end is placed under the surface of a soap solution in a shallow tank which sits on the stage of an overhead projector. That is really all there is to it. But the resultant raft of extremely regular bubbles, each about 0.5 mm in diameter, can be used to display and to study the behaviour of crystals in many different ways. Figure 1.18 shows an illustration from one of Sir Lawrence's papers[31] *on metal structures in which the behaviour of dislocations is illustrated.*

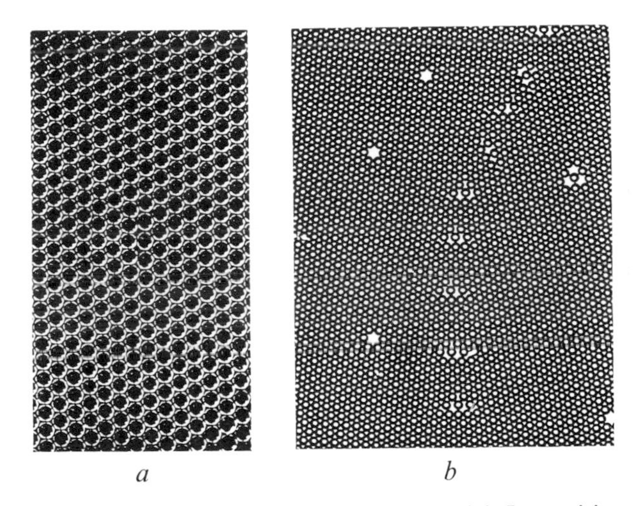

Figure 1.18 (*a*) A dislocation in the bubble model. Its position can be found by looking at the plate from the bottom left corner at a small angle of inclination. (From Bragg W L and Nye J F 1947 *Proc. R. Soc.* A **190**, figure 6a, by permission of the Royal Society.) (*b*) A row of dislocations along a boundary between two crystals which are nearly in the same orientation. (From Lomer W M and Nye J F 1952 *Proc. R. Soc.* A **212**, figure 3, by permission of the Royal Society.)

Two especially simple demonstrations that I remember very vividly for both simplicity and effectiveness were performed by Eric Rogers of Princeton.

Demonstrations 1.21 and 1.22

The first of these demonstrations is very simple in concept but needs considerable skill and practice on the part of the demonstrator. It concerns the parabolic trajectory of an object projected upwards at an angle to the vertical. A slide of a parabolic curve is projected on to the screen, which must be low enough and accessible enough for the lecturer to stand close to it with his or her hand near to one end of the curve. The lecturer then throws a coin or ball and, with considerable practice, it can be made to follow closely the curve on the screen. This sounds almost too simple; but in practice is quite dramatic and makes a lasting impression on the audience.

The second concerns the acceleration of falling bodies. A piece of string has a number of weights tied to it at equal intervals along its length. The lecturer then stands on a chair or table close to the edge of the dais or stage and holds the upper end of the string as high as possible. The length is chosen so that the bottom end is just touching the floor. The end is released and a series of clicks is heard as each successive weight reaches the floor. Although very rapid, the human ear–brain system can detect the time intervals and it is abundantly clear that the time intervals become shorter as the weights that started off higher reach the floor.

A second string is then produced, again carrying a succession of weights. But now the spaces between them increase according to a square law; that is, if the first is a distance x *from the floor at the start, the second is* 4x *from the floor, the third* 9x, *etc. Now when the string is released it is quite clear that the clicks fall at exactly equal intervals of time.*

I am not sure of the originator of my next example, but it was certainly performed regularly by Julius Sumner Miller (1909–87) who worked both in the USA and in Australia, and became widely known in Australia through his scientific demonstrations during commercials on television and for his catch phrase 'Why is it so?'

Demonstration 1.23

A rod or strip of wood or metal, such as a piece of welding rod or a metre rule, is supported on the two forefingers of the demonstrator, which are held out horizontally, pointing towards the audience (see figure 1.19). The fingers are then moved slowly together. No matter where the fingers were on the bar at the start, they always end up together at the centre of the rod. The demonstration is simplicity itself to do, but involves the concepts of centre of gravity, the law of moments and the dependence of friction on the normal reaction in order to explain it.

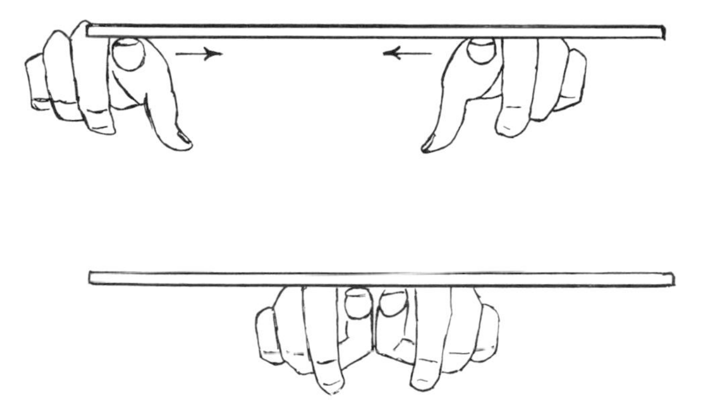

Figure 1.19 Sumner Miller's experiment on friction and normal reaction; wherever the fingers start they always finish together at the middle.

Another of Miller's demonstrations that is simple, but very effective, is to drive a paper drinking straw into a potato!

Demonstration 1.24

A good way to perform this is to invite a member of the audience to try to push a paper drinking straw into a raw potato, which turns out to be a seemingly impossible task. However, if the straw is gripped firmly about two inches from the end and squashed flat so that air is trapped in the straw when it meets the potato, it becomes possible, with a firm blow, to drive the straw quite a long way into the potato. Obviously the compressed air enables the straw to retain its cylindrical shape, and hence its strength and its cutting edge.

Miller's approach in his lectures, which were intended to popularise science, was to force his audience to think for themselves using techniques that sometimes came close to shock tactics. But it certainly worked and his lectures were enormously popular. He describes his approach in his book Millergrams[32]:

> *When people ask me to speak at some meeting or to lecture or deliver an address they sometimes say: 'Professor, how much time would you like?' To which I reply 'I do pretty good with a micro century' This sure gives them a quandary—a dilemma—a puzzle—which is, of course, just my intention. Cruel—wicked—calculating—all with the intention to make them* THINK.

1.7 THE ROYAL INSTITUTION OF GREAT BRITAIN

I make no apology for devoting a complete section to the work of the Royal Institution because it is one of the few places where lecture demonstration has been constantly encouraged for almost two centuries and it is a body about which I can speak from first-hand experience. It was founded by Count Rumford in 1799 and the first few years were somewhat disastrous; Garnett, Webster and Young all came and went in the first four years and it was Sir Humphry Davy, who first came to the RI in 1801 at the age of 23, who began to establish the reputation for excellence in lecturing. Indeed Davy's lectures were so popular among the fashionable society of the day that coachmen were asked to pick up and set down their passengers with the horses' heads towards Grafton Street in order to avoid congestion. This may well have been the first 'one-way street' in the world. Davy's success as a lecturer depended on several factors; he was young, handsome and charming and had a remarkable eloquence, and, at the same time, he was making remarkable scientific discoveries, particularly in relation to the isolation of chemical elements.

But it was Michael Faraday who really established the tradition of lecture demonstration. He originally arrived in 1813, aged 22, '. . . as assistant in the laboratory (at 25 shillings a week with two rooms in the attic and candles) . . .'[33] and became Superintendent of the House in 1821. In his earlier days he was much occupied assisting Davy, particularly with the work on the chemical elements. But in 1826 he started the Friday Evening Discourses and also the world-famous Christmas Lectures to young people which, except for a short break during World War I, have continued ever since. It is these two activities that became the breeding ground for some of the most effective demonstrations that have ever been produced.

One of the factors that has made lecture demonstrations at the RI so effective is its famous lecture theatre. It was built over some gardens next to the house in Albemarle Street and came into use in 1802; the basic design remains unchanged to the present day and it is one of the only theatres I know in which over 450 people can all see and hear extremely well wherever they are seated. It is semicircular, with the lecture bench at the centre, and very steeply raked; the lecturer can see every member of the audience and everyone has an uninterrupted view of the bench. The acoustics are remarkably good and extra amplification is rarely necessary.

Lecturers at the Friday Evening Discourses come from many parts of the world and it is worth recording one particular memorable evening that was reported by John Tyndall. It was a visit by Jules Antoine Lissajous (1822–80) to the Royal Institution in 1857.

Lissajous had exhibited his experiments on the combination of oscillations before the Societé d'Encouragement and before the French Emperor himself. Tyndall was intrigued by the demonstrations and planned to repeat them, but when Lissajous was consulted he offered to come to London himself. Tyndall reported the details of the lecture[34] and records that, at the beginning of the lecture, Lissajous

> . . . congratulated the audience on the presence of M.

Duboscq who took charge of his own electric lamp; this being the source of light made use of on this occasion.

Demonstration 1.25

The demonstration began with a 'sheaf of light' thrown from the lamp on to a mirror held in the lecturer's hand; when he moved the mirror quickly he could produce a ring of light on the ceiling and various other figures, thus illustrating the persistence of vision that is an essential feature of many of the remainder of the demonstrations. He then performed a number of experiments demonstrating what we now know as 'Lissajous' figures'.

For example he attached a small mirror to one prong of a tuning fork and arranged for a beam of light to be reflected from this mirror on to the hand-held mirror (see figure 1.20)

> *When a violin bow was drawn across the fork the image elongated itself to a line. By turning the mirror in the hand, the image upon the screen was resolved into a bright sinuous track many feet in length.*

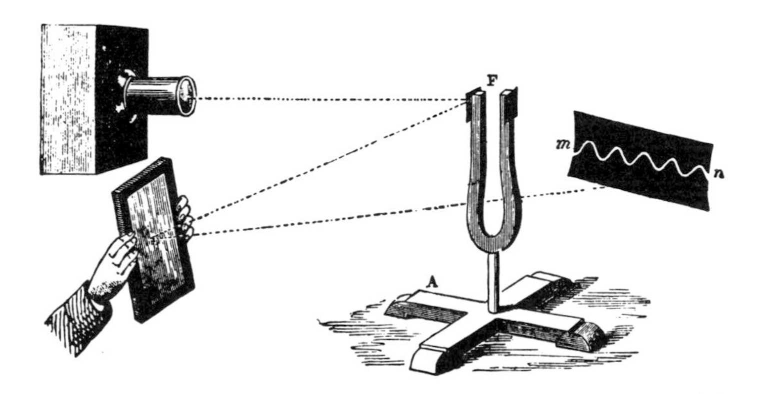

Figure 1.20 Lissajous' apparatus for drawing a graph of the vibration of the prongs of a tuning fork. (From Tyndall J 1875 *Sound* 3rd edn (London: Longman) p. 88, figure 22.)

He then went on to show various combinations of oscillations including the one most commonly associated with his name in which the light falls first on a mirror attached to one tuning fork and then to a mirror on a second fork placed so that the oscillation given to the light beam is at right angles to that given by the first.

He also demonstrated his 'phonoptomètre'. This was a microscope in which the objective was attached to a tuning fork so that it could be made to vibrate along a particular direction. If some other vibrating object, such as a string, was viewed through the microscope so that the two oscillations were at right angles, the frequency ratio between that of the fork and that of the object under test could be established with considerable precision.

> *At the conclusion of this beautiful series of experiments, which, thanks to the skill of those who performed them, were all successful, on the motion of Mr. Faraday, the thanks of the meeting were unanimously voted to M. M. Lissajous and Duboscq and communicated to those gentlemen by his Grace the President (The Duke of Northumberland).*

(I suspect that this was one of the earliest examples of an oscillograph-type demonstration in England. Nowadays this kind of demonstration is commonplace with the cathode-ray oscilloscope. But in recent times, just as Duboscq's lamp made the demonstration possible then, the laser has given it new life. The basic experiment with two tuning forks fitted with mirrors, or two vibrator units fitted with mirrors and driven by oscillators of variable frequency, is transformed into a remarkably beautiful display by using the beam from a small helium–neon laser reflected successively from the two mirrors. The brilliant red figures, when viewed in the dark, seem to acquire a three-dimensional character.)

One of the less tangible factors that tends to produce lectures of high quality at the Royal Institution is the inescapable realisation that many of the world's greatest scientists have lectured there; the atmosphere certainly produces the inspiration (and the necessary adrenalin) to promote good performances. At the Friday Evening Discourses the adrenalin flow is also stimulated by the tradition, instituted by Faraday after Wheatstone had disappeared ten minutes before he was due to lecture, of locking the lecturer in a small room at ten minutes to nine. This concentrates the mind wonderfully, and, on release to begin the discourse at nine o'clock, even the most blasé lecturer is conscious of the awesome task to be faced!

1.8 DEMONSTRATION IN DISTANCE-LEARNING PROJECTS

Demonstration can fulfil a particularly valuable function in distance-learning projects. The student is either following the course by correspondence or simply by studying the appropriate texts, and the occasional demonstration lecture over a television system, or by videotape, is one means of bringing the subject to life and ensuring that the students at least have some notion of the relationships between theory and experiment. There are many different kinds of distance-learning projects in existence, especially in Third-World countries. Lord Perry[35], former Vice-Chancellor of the UK Open University and Honorary Director of the United Nations University's International Centre for Distance Learning reports over 200 institutions in Western Europe and North America, and about a further 100 in Asia, Australia, Africa and Central and Southern America, actively engaged in distance-learning projects of one sort or another. In the West Indies the University makes use of satellite broadcasting; in Thailand, the Sukhothai Thammathirat Open University has well over 200 000 students; in India a very extensive system uses time assigned to higher education in the INSAT-1B satellite programme; and in the UK the Open University has its own BBC studio for the production of television programmes on the campus. It must be emphasised, however, that many of the projects do not use television very much because of the expense; in the UK, for example, it is said that television programmes cost at least seven times as much as radio programmes of the same length. There seem to be two main ways of using television; the simplest is one that I have seen in operation in the Republic of Ireland and in Australia. It consists in simply recording, with a single fixed camera, or even transmitting live along a land line, an actual lecture in progress either before a live audience or in an empty lecture room. The point of it is simply to provide lectures at a location where staff is short and without the necessity for the students or the lecturer to travel. It seems to work reasonably well, though with a single camera taking a wide angle shot of the lecture bench it is not always easy for the remote students to see exactly what is going on.

A more satisfactory technique is that used by the UK Open University, which is to make programmes illustrating particular points in a television studio. However, I have some reservations about the need to produce programmes that are as polished as those provided by the entertainment networks. I think there is a case for showing that experiments do not always work the first time and much could be learned by the students watching how the lecturer discovers the fault and corrects it—as they do in a live lecture. This point, together with other aspects of the use of television and video techniques, is discussed again in §2.6.

1.9 INTERACTIVE SCIENCE CENTRES

Science and technology museums in which some kind of interaction with the visitors is possible have been around for many years. The Children's Gallery in the South Kensington Science Museum with its buttons to press and handles to turn, for example, provided educational entertainment for many generations of children and I can still remember the excitement of my first visit some fifty years ago. The Palais de la Decouverte in Paris and the Evoluon in Eindhoven took the development a stage further; but the real catalyst in the extraordinarily rapid multiplication of truly interactive museums was Frank Oppenheimer who founded San Francisco's 'Exploratorium' in 1969.

The Exploratorium is a huge hangar-like building which houses about six hundred separate exhibits. Each exhibit is there to be used by the visitors rather than just to be seen. The categories are endless, and include sensory perception, heat, light, mechanics, aerodynamics, etc, etc. The following few examples give some idea of the range of topics and the level of scientific knowledge required to appreciate them: the 'Bernouilli Blower', which is a powerful air blower that will support a ten-inch-diameter ball; a vertical mirror arranged so that when the visitor places one hand at each side, the reflection of, say, the left hand, is mistaken for the right hand (confusion is caused when the real right hand is moved and the apparent right hand (the reflection) stays still); mimosa plants whose leaves close when touched; a Michelson interferometer whose fringes move when the beam on which the mirrors are mounted is touched; an enormously long and quite large diameter tube which produces a splendid echo effect.

In the last five or ten years the number of such interactive centres has increased enormously. The Questacon in Canberra opened in 1980; the Museum of Scientific Discovery in Harrisburg in 1982; Cardiff's Techniquest opened in 1986 and the Bristol Exploratory in 1987 (although pilot versions of the latter had been in operation for several years before that). A comprehensive account of the existing examples is given in a book by Stephen Pizzey[36].

How do they relate to lecture demonstration? I suppose they are complementary in the way that Armstrong hoped his discovery method would be at the end of the nineteenth century. But perhaps more closely related to the theme of this book are the 'lecturettes' that are being developed by some of these centres. For example, Canberra's Questacon has now grown into the National Science and Technology Centre and acts as a base for a travelling circus that tours round Australian schools giving short presentations about various topics accompanied by demonstrations and the possibility of follow-up experiments being done by the children themselves. The techniques of presenting these short programmes are closely similar to those needed for full-scale lecture demonstrations and, quite apart from their intrinsic value, obviously provide a splendid training ground for science communicators. I regard this whole movement as one of the most significant moves in the drive to popularise science.

1.10 THE USE OF DRAMA

A good demonstration lecture is a dramatic performance and there is no doubt that the lecturer must be at least a performer, if not an actor, in order to put over a demonstration lecture effectively. This may be one of the factors that leads some scientists to feel that there is something not quite respectable about such an activity; thankfully this attitude is beginning to be a thing of the past. Later in the book I shall from time to time draw parallels between a demonstration and a play; the same principles of variation in tension, of careful preparation and rehearsal, of clear diction and easily visible properties and actions all apply.

But there is also a place for dramatic presentations as a form of lecture demonstration and it is this aspect that I now want to discuss. A play is usually put on primarily to entertain, but, of course, there are often secondary purposes such as putting over a religious or moral message (as in mediaeval miracle plays), illustrating a social problem, or conveying a political philosophy. In a lecture demonstration the order of priorities is reversed and, though the entertainment element is still important, it is secondary to the main task of conveying the scientific principles. However, in recent years there have been several experiments in the integration of the two elements on a more equal footing[37].

One such experiment was the establishment of the Molecule Theatre by Lord and Lady Miles in 1967. Its intention is to put over a scientific principle and to stimulate interest in science for audiences in the 7–12 age group. The company is fully professional, with actors, producers, stage managers, designers, lighting and sound engineers and, of course writers. The script uses some kind of entertaining story as a vehicle and the action involves very large-scale demonstration experiments, performed by the actors, as part of the story. Also in the mid 1960s the project known as 'Theatre-in-education' was started, though it did not begin to tackle scientific topics until relatively recently. A small team of professionals, reminiscent of the travelling players of years gone by, travel from school to school and spend a full day working with pupils to develop a play. The value of this approach, of course, is that the children have to understand a piece of science quite clearly if it is to be incorporated satisfactorily into their own play.

A development in relating science and drama that is of special interest to me is the science and drama competition that was first held in 1981, and I propose to outline the main steps in its establishment. In 1973, a new university theatre was opened in Cardiff and a number of colleagues and I felt that it would be valuable for the science departments to become involved in the work of the theatre. What eventually emerged was 'Science Week'[38] which became a regular feature of the theatre programme once every two years. Lectures for all age groups from seven upwards, demonstrations, competitions and exhibitions of projects, etc, were held and proved to be extremely popular with local schools.

In 1981 John Beetlestone put forward the idea of a science and drama competition[37,39]. The idea was to invite schools to devise and perform a dramatic presentation which could be a straight play, a mime, a dance drama, a musical, etc, but must be based on a scientific theme. The science content could be an historical event, an aspect of the life of a scientist, a modern social issue arising from scientific developments or the exposition of a scientific principle. It should last 15 to 20 minutes, should use only a white or black back-drop, and the properties and elements of the set must be capable of being assembled on stage from scratch in five minutes. Groups of judges, including both science and drama experts, would visit schools to see the performances and the two best in each of the age groups 8–14 and 15–18 would be invited to perform in a final in the theatre during Science Week.

From the start it was made clear that not only the scientific (or technological) accuracy and the quality of the dramatic presentation would be considered, but also the effectiveness with which the two were integrated. In the early days there were one or two delightful presentations of entertaining dances or songs alternated with scientific exposition and these were not regarded by the judges as highly as performances in which the dramatic presentation was tightly geared to the scientific content.

The competition proved to be very popular and became an annual event. Then, in 1985, after discussions with the British Association for the Advancement of Science, it was decided to try it out on a national scale, with regional heats and a national final. In 1987, 80 schools took part, in five different regions.

I have included a brief account of this development because, although it is not demonstration in the usually understood sense of the word, it seems to me to be a very significant development that can fulfil some of the functions of the demonstration lecture or lesson. For example, later in this book (§3.9) I shall talk about audience participation as an important element in the demonstration lecture. Up to the age of about 12 or 13 children are quite uninhibited and are happy to join in spontaneously when invited in the middle of a presentation. Beyond that age it becomes a little more difficult. But the inhibitions that then appear do not seem to be so pronounced if prior arrangements are made, and if there is some kind of script. Role playing or a dramatic presentation can then be a powerful way of leading pupils to a deeper understanding of a scientific concept as a complementary process to that of straight demonstration.

Part 2

The Science behind the Art

2.1 WHAT IS A DEMONSTRATION?

2.1.1 Introduction

It may seem a little odd that we are already about one third of the way through the book and yet we are only just asking this question! But, so far, I have purposely taken a very broad view of what is meant by lecture demonstration and it is only now, when we begin to consider the more detailed and practical aspects, that we really need to worry about definitions. I suppose that by a demonstration one could mean any adjunct to a lecture that makes it more than a mere recital of words. Thus hand-waving in order to describe a spiral staircase, writing on a blackboard, or on the plastic sheet of an overhead projector, or the use of slides or films could all be described as demonstrations in this broad sense. The borderline between conventional visual aids and demonstrations is very ill-defined and so later on I shall say something about visual aids. But generally speaking a demonstration means the illustration of a point in a lecture or lesson by means of something other than conventional visual-aid apparatus.

I like to divide demonstrations into three categories:

(1) visual aids using non-conventional apparatus;
(2) analogue demonstrations;
(3) real experiments.

In the index to demonstrations I have indicated the category into which each demonstration falls. As illustrations I shall describe two examples from my own repertoire that fall into each of the three categories.

2.1.2 Unconventional visual aids

A good example of a visual-aid type of demonstration is the striped string that I use to help audiences to understand the concept of the limit of resolution in an optical system.

Demonstration 2.1

There are two stages in the production of an image by an optical system, or indeed by any other kind of imaging system, such as radar, ultrasonic scanning, electron microscopy, etc. In the first stage the radiation is scattered by the object and, in the process, information about the object is encoded. In the second stage the coded information is sorted out to form an image. The two stages can be called scattering and recombination, or diffraction and focusing, or coding and decoding. The information is encoded in terms of phase differences, and to illustrate how this arises, and how it affects the detail in the final image, we can use the striped string model[40]*.*

The apparatus needed is a piece of white string with black stripes painted on it to represent the peaks and troughs of a set of waves and a hook at either end, together with a strip of metal punched with regularly spaced holes along its length. The strip represents an object to be imaged and the hooks are attached to two holes, one at each end of the strip. The strip is held vertically in the left hand and the loop of string is passed over

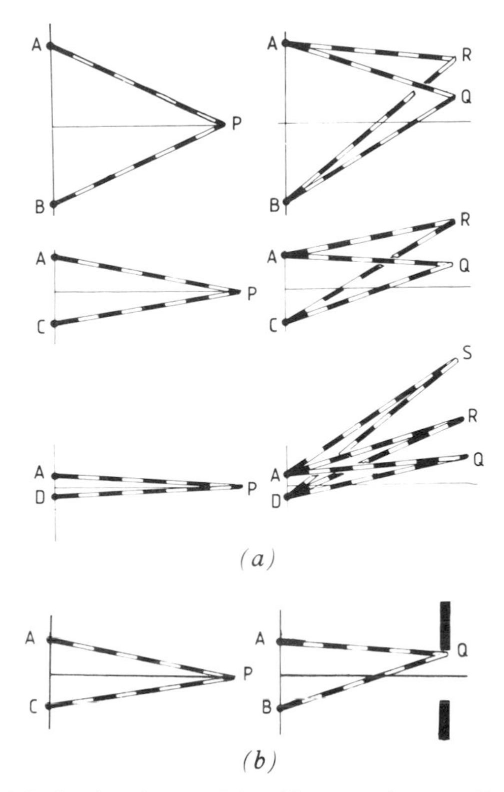

Figure 2.1 (*a*) Striped-string model to illustrate the encoding of information in the form of phase differences between waves scattered from two different points on an object with three different spacings. (*b*) Striped-string model to illustrate the restricting effect of an aperture on the range of possible phase changes for the two points an intermediate distance apart in (*a*). (From Taylor C A 1987 *Diffraction* (Bristol: Adam Hilger) pp. 36 and 37, figures 4.2 and 4.3.)

the right forefinger to represent a point in the scattered radiation from the object (figure 2.1(a), top pair). If the right forefinger is raised and lowered the change in relative-phase between the waves scattered from the two ends of the strip can be seen. If the points of attachment to the strip are moved closer together (middle pair) then the phase difference changes more slowly with the position of the forefinger. If the points of attachment are closer together than the wavelength (figure 2.1(a), bottom pair), then there is very little phase change even if the forefinger is moved from one extreme to the other. This demonstrates clearly that, if an object is much smaller than the wavelength of the radiation in use, no significant phase differences can be created. So no information about the object can be encoded, and hence, however good the subsequent image-forming system is, it will never be able to form an image; the necessary information is simply not there. This condition is known as the wavelength limit of resolution.

The same apparatus can be used to demonstrate how, even if the object is large compared with the wavelength, no image can be formed if the aperture is too small. Figure 2.1(b) shows the points of attachment at the ends of the strip, as for figure 2.1(a) middle pair, but the right forefinger can only move over a small range, corresponding to the restricted aperture, and so very little change in the relative phase can be detected. This condition is known as the instrumental limit of resolution.

A second example of a visual aid is the use of wire models to illustrate interference effects in polarised light. The basic idea was originally suggested by Spottiswoode[41] but it has been modified by various lecturers since then. The version that I use is described below.

Demonstration 2.2

Two lengths of stiff wire (I use 2 mm welding rod) are bent into the shape of sine waves with an amplitude of about 50 mm. They are connected together at their points of zero displacement by means of small rubber bands so that, by rotation about these zero points, they can be made to represent sine waves whose planes of polarisation can have any desired angle between them (see figure 2.2). Three discs are mounted on a baseboard so that they can be rotated in their own plane; the first and third carry a single slot a little longer than twice the amplitude of the waves and wide enough to give plenty of clearance when the wires are passed through. The middle disc carries two slots at right angles to each other. With the slot in the first disc vertical, the two wire waves, in phase with each other, are passed through (figure 2.2(a)); this represents the transmission of plane polarised radiation by a polariser. The middle disc is set so that its slots each make an angle of 45° with the vertical and the wire waves are rotated about their null points so that one can pass through each slot of the middle disc (figure 2.2(b)); the wire waves now represent two components of the initial wave, polarised at right angles to each other.

With the final slit set horizontally the only components that can pass through are the horizontal ones, and when the wire waves are rotated to be horizontal they will be found to be precisely out of phase (figure 2.2(c)). Hence, in the case of a real analyser, no radiation would emerge. If, however, the two components created at the middle disc travel at different speeds, they arrive at the analyser with a phase difference that depends on these relative speeds and the distance travelled. This is exactly what happens when a beam of plane polarised light falls on a plate of doubly refracting material (mica for example); the two slots in the middle disc represent the ordinary and extraordinary vibration directions in the mica. The phase difference will depend on the thickness of the mica, on the difference in refractive indices for the ordinary and extraordinary rays, and, since this in turn is different for different wavelengths, the result is that the field becomes coloured. The same model can be used to help an audience to think through many other examples of interference in plane polarised light.

2.1.3 Analogue demonstrations

My second category of demonstrations is that of analogues. In this case the demonstration uses a phenomenon whose behaviour is sufficiently similar to that being discussed to make it valuable as an instructional aid. My first example in that category would be the use of optical analogues to illustrate x-ray diffraction. The problem here is that the phenomenon being discussed, in this case x-ray diffraction, is not easy to demonstrate directly because x-rays are both invisible and dangerous. However, with a suitable change of relative scale of the object and of its nature, many of the phenomena of x-ray diffraction can be demonstrated to an audience. Indeed the analogy is so close that this technique has been widely used as a research tool, especially in the period before digital computers came into use[42], and, indeed, is still in use in certain problems for which even a powerful computer is inadequate[43].

(a)

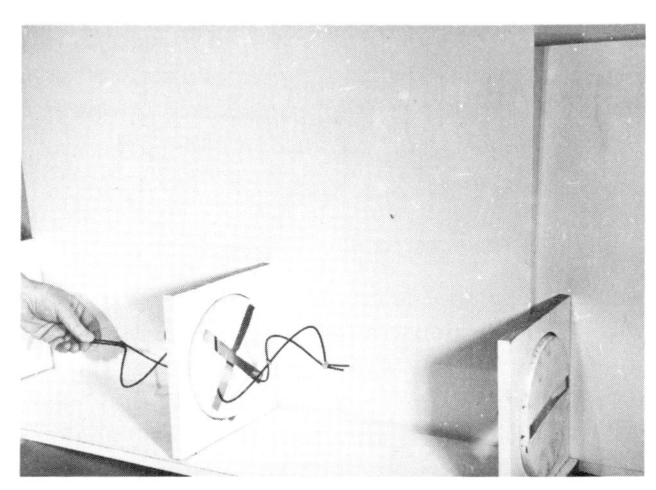

(b)

(c)

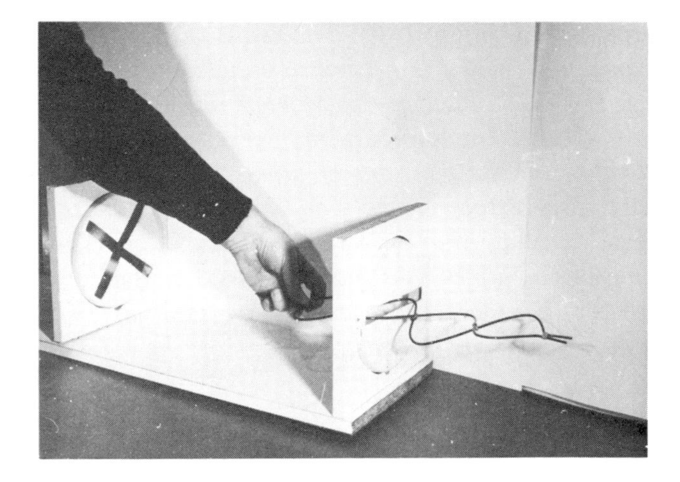

Figure 2.2 Adaptation of Spottiswoode's idea of wire models to illustrate interference in polarised light. The two wire sine waves are connected together at the zero displacement points by small rubber bands. (*a*) The polariser produces plane polarised waves. (*b*) Two components, each at 45° to the plane transmitted by the polariser, travel through the object (e.g. a sheet of mica) vibrating in the fast and slow directions. (*c*) At the analyser the horizontal components are exactly out of step if the object is isotropic, but their phase difference will vary with the thickness and with the refractive indices for the fast and slow directions if the object is birefringent.

Demonstration 2.3

Many people think that the real difference between x-ray and optical diffraction lies in the fact that optical diffraction is usually studied with planar diffracting objects (masks) and x-ray diffraction with three-dimensional objects (atoms arranged in a crystalline array). However, the significant difference lies in the relative size of the wavelength compared with the size of the object. Thus in a typical x-ray diffraction experiment the wavelength (perhaps 1.54×10^{-10} m) is virtually the same as the separation of the centres of carbon atoms in an organic molecule; as a result, the significant angles of diffraction are *very large indeed—up to 180°. The result is that quite complicated camera geometry is needed to enable the patterns to be recorded. But we can ignore this problem and it can be shown*[44] *that the optical diffraction pattern of a set of holes punched in card to represent the positions of atoms in a two-dimensional projection of a certain crystal can closely resemble the x-ray pattern produced by that crystal (see figure 2.3). Usually the optical mask has dimensions that are large compared with the wavelength of light, and hence the angles of diffraction involved are small and the awkward geometry of the x-ray case is avoided.*

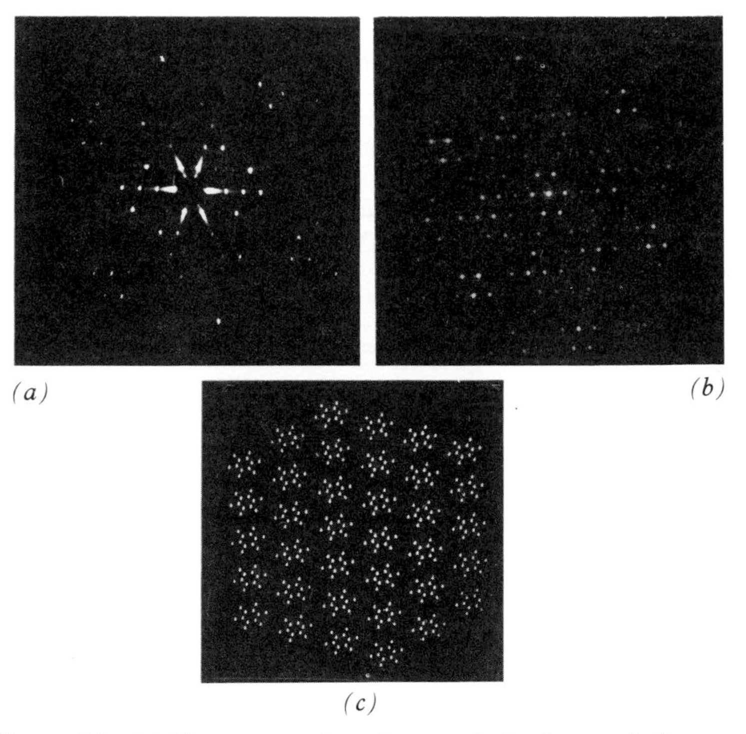

Figure 2.3 (*a*) X-ray precession photograph for hexamethylbenzene. (*b*) Model of a possible structure for hexamethylbenzene. (*c*) Optical diffraction pattern of (*b*). (From Taylor C A 1987 *Diffraction* (Bristol: Adam Hilger) p. 47, figure 5.2.)

If a long optical bench is available it is not difficult to set up apparatus to demonstrate this to an audience[45]. *But I have a simpler arrangement that is quite easy to make; the only expensive item is a 0.5 or 1 mW helium–neon laser. The laser is mounted horizontally at one end of a long bench or table (e.g. about 3 m) and there is a ground glass screen in a wooden holder at the other end. The remainder of the items (lenses, slits, mask holders, etc) are mounted in home-made sheet brass holders attached to 5 mm brass-rod stems which, in turn, fit into holes in cylinders of duralumin about 50 mm in diameter and 60 mm long, and can be set at a given height by means of a screw (see figure 2.4). This arrangement can be used in different configurations to display many kinds of diffraction phenomena (see also §2.6) but to demonstrate the analogy with x-ray diffraction the configuration already shown in figure 2.4, and diagrammatically in figure 2.5, is used. The beam from the laser source illuminates the object, which is a mask produced photographically to represent the arrangement of atoms in a projection of a crystal structure; it is placed about 1.5 m from the laser (in order to take advantage of the slight expansion of the beam). Without any of the other components in position, the diffraction pattern of the mask appears on the screen placed a further 1.5 m beyond the mask. If large audiences are involved the image can be picked up from the ground glass screen with a simple black and white television camera and displayed on a large television screen (see Demonstration 2.10, p. 89). A is a lens of focal length about 100 mm and can be used to focus an enlarged image of the mask on the screen. (This corresponds to the recombination of the scattered waves mentioned in the discussion of Demonstration 2.1, p. 60, an operation that cannot be performed with x-rays because no x-ray lens is possible.)*

The boundaries between the categories are somewhat diffuse and it is not always easy to distinguish them clearly. However, here is another example that fits fairly clearly in the analogue category; this time it is very simple, but nevertheless very effective.

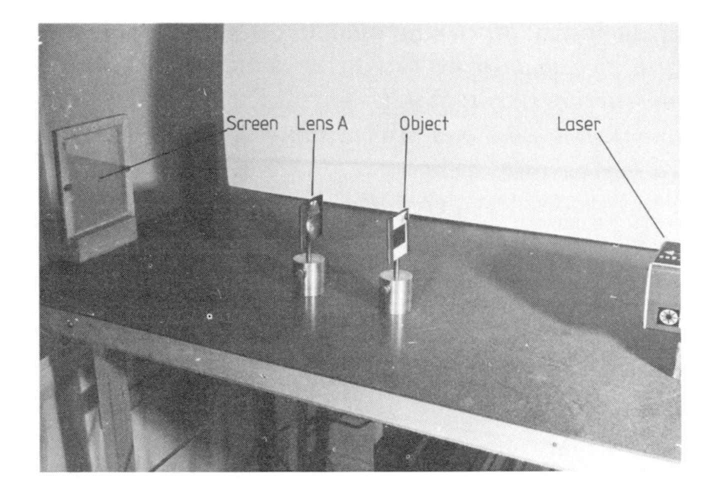

Figure 2.4 Portable diffractometer.

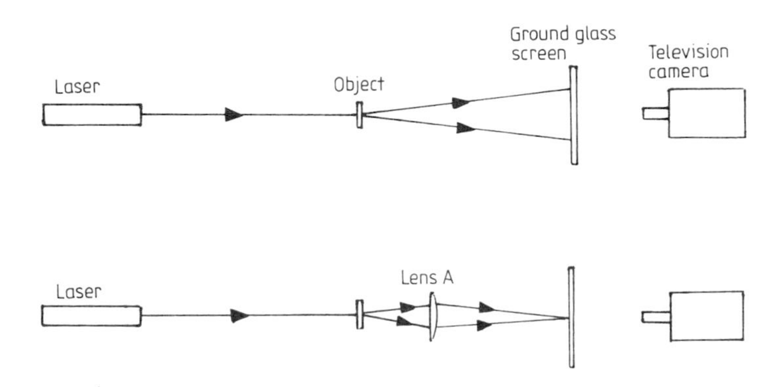

Figure 2.5 Portable diffractometer (diagrammatic).

Demonstration 2.4

In the study of sound, it is very common to illustrate modes of vibration by means of transverse standing waves on a string or rope (the experiment is sometimes called Melde's experiment, though he did it in a somewhat different form). But students often find it difficult to relate these transverse modes to the longitudinal modes in a pipe. The toy known as the 'Slinky spring', or a helical coil of springy wire, provides an excellent analogue demonstration. The spring is stretched between the demonstrator's hands and allowed to lie horizontally on a smooth table. A sharp movement of one hand sends a clearly visible compression along the spring, but simultaneous in and out movements of both hands can build up standing waves with clearly visible nodes. With a little practice modes with one, three and five nodes can be shown (see figure 2.6).

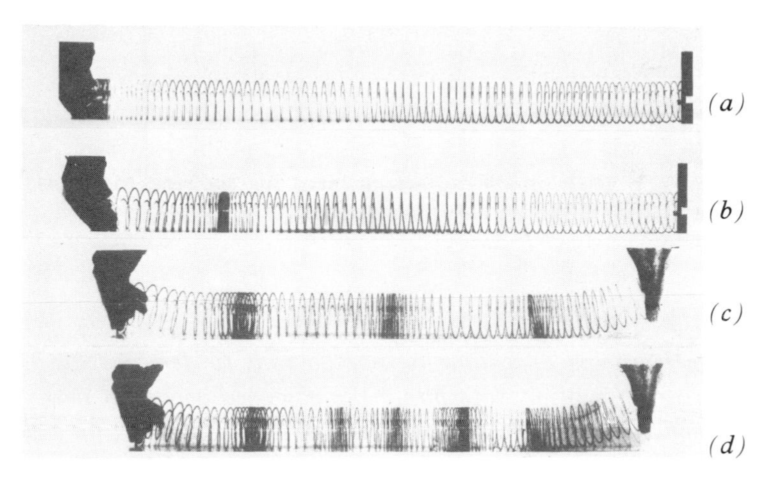

Figure 2.6 Longitudinal waves on a helical spring. (*a*) Spring at rest. (*b*) Pulse travelling along from the left. (*c*) Standing waves showing three nodes. (*d*) Standing waves showing five nodes.

2.1.4 *Real experiments*

I shall now give two examples of demonstrations that are in my third category—those in which the actual phenomenon being discussed is portrayed. Both the examples I have chosen are relatively simple, but elsewhere in the book descriptions of much more complicated examples are given.

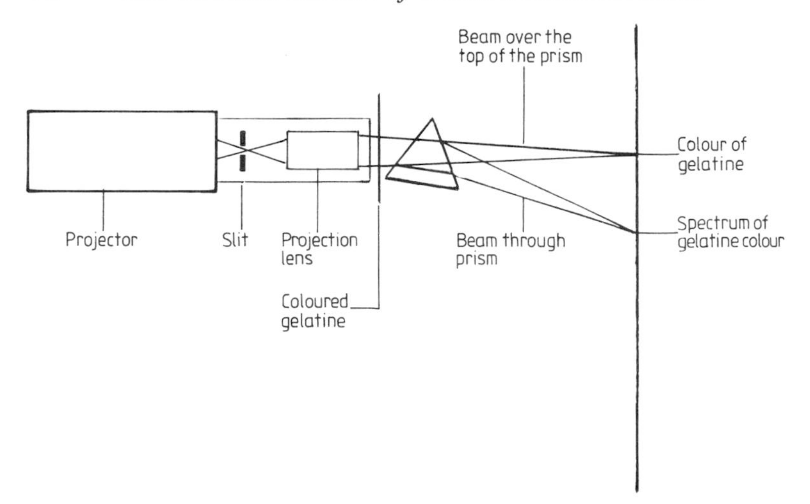

Figure 2.7 Diagram of the arrangement for showing the spectrum transmitted by a coloured transparent object.

Demonstration 2.5

When lecturing on colour it is important to demonstrate how selective absorption determines the observed colour of many of the objects around us. A simple way to do this is to set up a slide projector with a slit (say about 1 mm wide) in the normal slide position and to position a prism in front of the projection lens in order to produce a spectrum (see figure 2.7). (It is, of course, important to make sure that the spectrum is as pure as possible by focusing the objective to give a sharp top and bottom edge to the spectrum and by setting the prism at the position of minimum deviation.) The height of the prism is adjusted so that some of the light from the projector passes over the top and so a white image of the slit appears on the screen directly ahead of the projector; the spectrum appears at one side either on the same screen or on a second one. Pieces of coloured transparent plastic, or of glass, or tanks containing coloured dyes can then be placed in front of the objective lens. The colour of the object can then be seen on the screen ahead of the projector and its component spectrum can be seen on the screen to one side. A particularly striking effect is obtained if a magenta colour filter (e.g. a magenta 'gelly' of the type used in stage lighting) is used. The green portion of the spectrum is neatly cut out leaving the remainder intact: this provides an excellent introduction to the idea of negative colours; magenta sometimes being described as 'minus green'. Incidentally, tanks with plane parallel sides to show absorption by liquids can be very expensive; I have found that the clear plastic jugs that are sold by supermarkets to contain one litre cardboard packs of fruit juice work sufficiently well for demonstration purposes and are very cheap.

Finally I include details of a simple but effective demonstration of the idea of action and reaction. Of course if a linear air track is available the demonstration can be made more quantitative, but the version described here involves very little expense.

Demonstration 2.6

A length of toy railway line, say about 0.5 m long, and a toy truck that will run smoothly on the track are needed. A smooth piece of wood or plastic is mounted obliquely in the truck (see figure 2.8). The next essential is one of the jumping toys that can be bought very cheaply; they are often in the shape of a frog or some sort of insect. The 'works' consist of a spring and a rubber sucker and, if desired, can easily be removed from the toy and used on their own. The spring is compressed and the sucker pressed on to the oblique attachment to the truck; after a few seconds the sucker works loose and the toy is projected in one direction and the truck moves off down the line in the other direction.

Now that we have arrived at an understanding of what basic types of demonstration can be distinguished—at least by case histories if not by verbal definition—we need to consider some of the basic problems of information transfer that are solved by the use of demonstrations.

Figure 2.8 'Jumping frog' toy with the frog removed and mounted on a toy railway truck. The spring is compressed and held for a few seconds by the suction pad; when it releases itself the truck recoils along the track.

2.2 PROBLEMS OF INFORMATION TRANSFER

2.2.1 Information

The word 'information' is used in many different contexts; it is used in everyday life to denote facts such as times of trains, routes from A to B or where certain commodities can be bought. In the world of libraries it relates to reference books such as encyclopaedias, dictionaries, who's who's and to the indexing systems that enable books to be located. In the world of computers it can become quantified in terms of numbers of words, or of kilobytes, and it might, for example, refer to the contents of a database (that is a computerised file of data from which items can be called up to the video screen at will). In the world of telecommunications it relates to the quantity of data that is being transmitted from A to B, but it also relates to the quality of the information. For example the early 405-line television system does not provide as high a quality of picture as the present 625-line system, and that in turn can be improved on by more recent ultra-high definition systems; but as the quality goes up, so the information needed for transmission measured in bits, or bytes per second, goes up.

In all these examples communication of some kind is involved; information locked away in a book or a computer file is of no use unless it can be extracted by the originator when required, or passed on to someone else. Sir David Attenborough, in his book *Life on Earth*[46], makes the very interesting point that the genetic information stored in DNA can be completely lost when a species becomes extinct. But he then goes on to say that libraries can be thought of as immense, communal brains that can memorise far more information than could be stored in a single human brain. He goes on to say that libraries

> ... can be seen as extra-corporeal DNA, adjuncts to our genetical inheritance as important and influential in determining the way we behave as the chromosomes in our tissues are in determining the physical shape of our bodies. ... Cut off from our libraries and all they represent and marooned on a desert island, any one of us would be quickly reduced to the life of a hunter-gatherer. ... Man's passion to communicate and to receive communications seems as central to his success as the fin was to the fish or the feather to the birds.

2.2.2 Usability of information

In a paper that I wrote[47] for the Bangalore Conference on Science and Technology Education and Future Human Needs I introduced the concept of 'usability' of information. I likened it to the consequence of the second law of thermodynamics, which tells us that we can have enormous amounts of energy stored, for example in a thermal reservoir, but, without another reservoir at a lower temperature, the energy is unusable. In a roughly parallel way we can have enormous amounts of information stored in various kinds of reservoirs but it may not be *usable* information:

> Why should information be unusable? It could be lost

> because it has not been efficiently labelled. A book can be lost for a long time if it is replaced on an incorrect shelf in a library, or if the original cataloguer misunderstood the subject; a file can be lost if it is incorrectly labelled or placed in the wrong cabinet; data in a computer information bank can be irretrievably lost if the key words are not skilfully selected.

But, in addition to the idea of usability, we need to think about the efficiency of transfer; the very act of communicating the information, or of information transfer from one store to another, can introduce distortion of various kinds. An obvious example would be an attempt to hold a conversation over a bad telephone line, or the reproduction of tape-recorded information by a very poor amplifying system which introduces so much distortion that the message cannot be understood. In Demonstration 2.1, p. 60, we talked about the limitations imposed by optical systems on visual information, and in Demonstration 2.10, p. 89, we shall see that not only is it possible to obliterate visual information but also to actually falsify it. In a microscope, for example, if the aperture is too small false detail may be introduced into an image (see figure 2.13).

The analogue to be used here is that when I lecture I am acting as an information-transmission system and can affect the quality of information transmitted all too easily. Obviously, if I have misunderstood the phenomenon about which I am talking I will not be able to convey accurate information to my audience; if I understand my subject but am not able to explain it clearly to my audience then the efficiency of transfer may be very low; but even if I know my subject and can explain it clearly, the audience can only perceive the topic through my brain, and that acts as a kind of aperture which may modify, restrict or even falsify the information. This can be a good thing: it means that the lecturer's personality is imposed on the subject matter and three different lecturers of equal expertise in the subject and in the art of lecturing would produce three quite different lectures on the same topic. But, equally, it could be a bad thing and could lead to the audience picking up a completely false picture of the subject and of its implications; the lecturer is in an enormously powerful position of influence, and therefore has a responsibility to make the efficiency and accuracy of information transfer to the audience as high as possible.

In my view this is one of the important justifications for using demonstrations. Audiences can relatively easily misunderstand words. A momentary distraction may mean that a listener misses an all-important qualifying system, or even a negative, so misunderstanding a whole sequence of ideas. If the point is well illustrated by a demonstration the chance of imperfection in the information transfer is greatly reduced.

2.3 AUDIENCE PSYCHOLOGY

2.3.1 Introduction

It is important to recognise that, just as in the live theatre, there is a complex interaction between a lecturer and the audience. A scientific lecture is just as much a dramatic performance as is a play and the same lecture with the same demonstrations can turn out to be totally different with a different audience. I have given some of my lectures hundreds of times and am often asked whether I am in danger of becoming bored with the topics. The answer is that the audience makes this an impossibility; until you have actually begun to speak you have no idea what the audience reaction will be and, once started, you are too busy thinking on your feet about how to adapt to the particular audience reaction to be bored!

2.3.2 Conditioning an audience

There is a well worn adage that in any lecture or lesson you must 'Tell 'em what you're going to tell 'em, then tell 'em, and then tell 'em what you've told 'em' and I can find no fault with that. It is certainly good advice to start with. But of course there is much more to say. I remember Dobson, of the wartime broadcasting team of Dobson and Young, telling a story about an experience he had when lecturing to an audience of soldiers (a captive audience) on classical music. He had 'softened up' the rather reluctant listeners for about ten minutes when his lecture was interrupted by a sergeant who removed half the audience for other duties and marched in a replacement set of even more unwilling victims! Imagine the problem of coping with the half who were just beginning to feel that the lecture might not be such an ordeal as they expected and the new half completely cold.

Warming up the audience is obviously important and I find that it is not a good idea to introduce any points that are vital to the understanding of the rest of the lecture during the first few minutes. The audience need to get used to the sound of your voice and your style and to generally 'settle down'. I very often use a few slides that are relevant to the lecture but not an essential part of the logic; for example in a lecture on science and music I usually begin with a few examples of the way cartoonists have treated music and musicians. A little humour at the beginning can work wonders in relaxing the audience. But, of course, there are dangers. I once made the mistake of trying this approach when I was lecturing abroad to an audience with a totally different cultural background and sense of humour, with disastrous results.

2.3.3 Varying the tension

Having begun on the 'meat' of the lecture it is important to adopt the technique used by dramatists of varying the tension in order to retain attention. Build up to a climax of some sort and then, perhaps every ten minutes or so, release the tension with a joke or a frivolous addition to a demonstration and then start to build again. It is interesting to discover how often the points made in a lecture just *after* a lighthearted interlude are the ones that are remembered. Again, quoting my music lectures, I sometimes illustrate the variable frequency vibrator as an element in a musical instrument by playing the musical saw, and that is certainly remembered by audiences together with the more serious points made in the same part of the lecture.

2.3.4 Some properties of the brain

It is very important for a lecturer to be aware of some of the remarkable and curious properties of the human brain. Those whose responsibility it is to interview witnesses to an accident or crime are well aware of the fact that a number of quite impartial witnesses will give quite different accounts of the same event. Most people are aware of the fact that, if you view a film or television programme for a second or third time you will often see things that you did not notice at all the first time, or you may even realise that something you thought you saw the first time did not in fact occur at all. How, then, can we be sure that our lecture audiences are observing accurately?

Obviously one solution would be to do every demonstration several times; but this may not be practicable either because of the time taken during the lecture, or because the experiment takes a long time to reset. How can we get round this problem? My solution is really a version of the old adage mentioned above. Tell them what they are going to see or hear and what to notice particularly; then do the demonstration; then remind them of what they should have seen or heard.

I have a favourite demonstration that illustrates how powerful this technique can be.

Demonstration 2.7

To do this demonstration you need a recording of a piece of speech , preferably of a sentence that is slightly unexpected and not directly connected with the topic under discussion. The recording is made with some kind of defect or distortion so that the combination of the defect and the unexpected nature means that the audience do not understand it. If you then say the sentence clearly to the audience a couple of times and finally play the recording again the audience will understand it perfectly and will often refuse to believe that it is indeed the recording that they could not understand on first hearing. A useful form of distortion is to speed up the speech very considerably (machines, both analogue and digital, exist which can do this without altering the pitch). I came across an excellent recording for the purpose quite accidentally. A sentence had been synthesised for me in order to demonstrate the successive stages in the process; I found that I could understand the sentence perfectly even before the process was complete but was puzzled by the blank looks on the faces of my audience when I said that it was already understandable. Then, of course, I realised that I could understand it because I knew the sentence; the murmur of amazement that greets the second playing after the audience has been told the sentence leaves one in no doubt that the demonstration has worked!

I have sometimes referred to the process of telling the audience the sentence as 'brainwashing'. And I do not think that that is too powerful a word. It arises from one of the most fundamental properties of the brain; the storage of items in the memory and the almost instant recognition that what is being experienced at the moment is already in the memory. It is the process that enables us to pick out speech we wish to hear from the background noise of the cocktail party, or that enables us to recognise a line drawing as a representation of a three-dimensional scene.

2.4 VISUAL AIDS AND EDUCATIONAL TECHNOLOGY

2.4.1 Introduction

I suppose that the earliest examples of a visual aid in the

ordinary sense of the word are the large scale maps or charts that have hung on the walls of school rooms almost since schools began. Blackboards, too, must have an honoured place. But I remember quite clearly that when I was at primary school in the late 1920s, there was great excitement when a geography lesson was embellished by a display of rather dim sepia-tinted slides projected by an enormous magic lantern that required three boys and the caretaker to carry it into the room.

I also remember that my secondary school possessed an epidiascope, which not only projected $3\frac{1}{4}'' \times 3\frac{1}{4}''$ slides, but was also capable of projecting quite an acceptable image of a picture on opaque paper or from a book. And our physics lessons were enlivened by the projection of 16 mm sound films. At that time, of course, the use of such aids depended chiefly on the initiative of particular teachers. Indeed I believe that the cine projector was the personal property of the physics master. These simple and effective aids remained almost unchallenged until the 1960s when there came the flowering of educational technology departments in universities and of resource centres for schools. Suddenly visual aids became educational technology and educational technology became big business. Tickton[48] defines educational technology as

> . . . a systematic way of designing, carrying out, and evaluating the total process of learning and teaching in terms of specific objectives, based on research in human learning and communication, and employing a combination of human and non-human resources to bring about more effective instruction.

This is a formidable definition, but I think the two words that specifically need to be emphasised are 'evaluating' and 'effective'. In the 'bad old days' teachers used slides, recordings, charts, etc, without trying to evaluate their effectiveness; they depended on gut feelings to enable them to decide whether a particular type of presentation was worthwhile.

While not wishing to pour scorn on all the ideas of

educational technology, it does seem to me that, as in so many other areas, the enthusiasts were carried away to extremes which tended to bring the whole subject into disfavour. The real problem, I am quite sure, is one that crops up in many places in this book; it is the confusion between using a technique because it makes the transfer of information, or the learning process, more effective, and using the technique for its own sake; the confusion of the message with the medium.

The point is brought home very clearly by Postlethwaite and Mercer[49] in a discussion of 'multi-media' approaches to teaching science:

> One expects the approach to involve a learning centre or study room equipped with electronic devices which make adjustments for individual differences and cause students to achieve at a high level and at a very rapid rate. The teacher is replaced and students proceed independently with little or no contact with other human beings. This image is unfortunate for, while some of these features are possible and important, the use of media does not necessarily replace the teacher, make education less expensive, or result in greater learning. The confusion arises primarily from a rather superficial understanding of what media can and will do and how the rôle of the instructor and student may change when these materials are being used.
>
> The definition of a medium as 'an agency, such as a person, object, or quality, by means of which something is accomplished, conveyed, or transferred' is helpful for placing the relationship of teacher–medium–student in proper perspective. In other words the medium is a means to an end and not an end in itself.

Now, in the late 1980s we seem to have reached a more balanced position and, in spite of the availability of computers, video recorders and video-discs, slides, audio tapes and overhead projectors seem to have established and retained a firm and useful place in the teacher's armoury of aids.

2.4.2 Slides

The great attractions of slides are that they are relatively cheap, that with modern colour films they can be quite stunningly beautiful, that each slide represents a single point so that they can be assembled to order to illustrate the teacher's own sequence of ideas, and that it is quick and easy to jump back and forth within a set of slides. The difficulty with film or video presentation (unless one can afford the latest computer–video-disc combinations) is that you have to use the sequence ordained by the producers and that it is much more difficult to jump about within the sequence.

2.4.3 Tape and slide combinations

A very powerful combination is the use of slides together with audio tapes, and, indeed, some advertising agencies, who should know about effective communication, use several slide projectors controlled by signals on one track of the audio tape in preference to film or video. In the field of science education one of the most effective users of tape and slide combinations that I know of is Mark Boulton of the International Centre for Conservation Education, who uses it in putting over ideas on the conservation of nature and natural resources to schools. But he also gives an important warning[50] against undervaluing the intelligence of the audience and points out that the use of slides does not *automatically make any talk interesting*. He says

> The most effective presentations are often those where the visuals form a 'structured background' to a talk which *would be interesting in its own right—even without the pictures.*

2.4.4 The overhead projector

Overhead projectors have become almost standard equipment in most lecture rooms. When they were first introduced they were presented as a substitute for the blackboard with the added advantages that the lecturer could remain facing the class (how well I remember one distinguished lecturer from my own student days whose voice was very quiet anyway, and since he spent most of his time facing the blackboard and covering it with mathematics we could scarcely hear at all!) and also retain a permanent record of what had been written. Nowadays the great additional point is that it is so easy to prepare transparencies in advance that many people use them in place of slides. Modern copying machines will produce good transparencies on suitable transparent film and, with a computer graphics facility some remarkably useful charts, graphs, etc, can be produced quickly and easily. And, of course, with an erasable pen, notes can be added during the lecture and removed later.

In very recent times two additions to the overhead projector have become available and, although still relatively expensive, their price is coming down and they should prove enormously valuable for demonstration lectures. They both involve liquid-crystal devices. A glass sandwich is placed on the overhead projector table and is fed with electrical signals. In one case the result is the display of the contents of a computer video display and in the other it is the screen of a cathode-ray oscilloscope. In both the relevant patterns appear in black on white and, of course, can be projected up to a size that permits them to be seen by a very large audience.

I also find the overhead projector invaluable in other ways for demonstrations, as in the following examples.

Demonstration 2.8

In talking about crystal structures and of the way in which crystals grow I place a large watchglass on the table of the overhead projector and then add steels balls (such as are used in ball-bearings). If the balls are of miscellaneous sizes, the patterns produced are quite irregular; but if all the balls are identical beautiful two-dimensional representations of close-packed arrangements can be produced. With care, dislocations can be produced and grain boundaries can be demonstrated. The idea of slipping of one plane over another is easy to show and the occurrence of one or two larger balls inhibits slip and gives some indication of how the addition of a small proportion of larger atoms can harden an alloy considerably (e.g. the addition of copper to aluminium) (see figure 2.9).

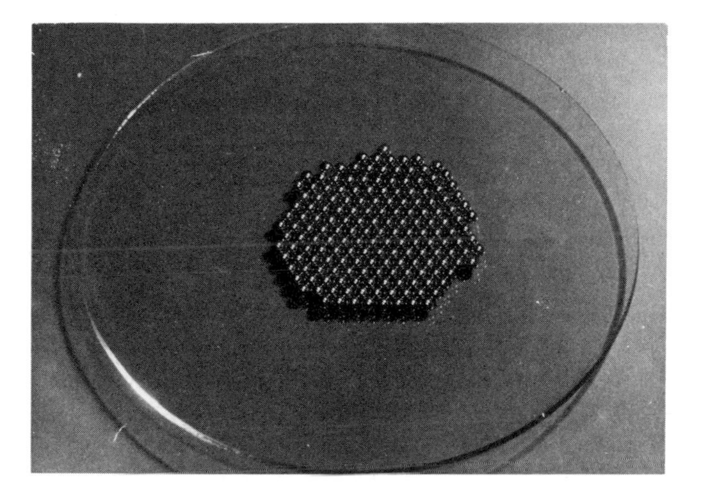

Figure 2.9 Steel balls on a watch glass illustrating the close packing of atoms in a metal-like structure.

Demonstration 2.9

The overhead projector is useful in demonstrations about colour. I use a series of small plastic petri dishes, about 50 mm in diameter, which can be filled with dye solutions to show, for example, why it is preferable to use magenta, yellow and cyan as the primary subtractive colours rather than the red, blue and yellow that is still being taught by many art teachers. I use greatly diluted photographic dyes, and with only 5–10 ml of the diluted solutions a large audience can see the results easily. Mixing yellow and blue, for example, gives a greyish colour, rather than the green that many people predict. The reason is that my carefully chosen dyes are such that the yellow removes only the blue component, but the blue removes both red and green; when the two are mixed all three components are removed. But if magenta (minus green), yellow (minus blue) and cyan (minus red) are used, very vivid red, blue and green can be produced by mixing in pairs. (See also Demonstration 3.9, p. 136)

Educational technology, according to the definition used earlier, should include the use of film, television, microcomputers and video techniques, but it seemed better to devote special sections to them.

2.5 USE OF FILM AND VIDEO RECORDING

2.5.1 *Introduction*

There are many excellent handbooks on the techniques of film and video making and I do not propose to discuss them here; my purpose is to discuss the pros and cons of using these media in place of demonstrations. I have already indicated in the last section that one of the problems in using either film or video recording (from here on in this section I shall use the term film to include both) is that the sequence of ideas is fixed and it is not easy to drop in at a particular point in order to use a small section.

2.5.2 *Justification of the use of film*

In spite of that difficulty there are obviously circumstances when the use of these media seems to me to be completely justified. For example, it is obvious that there are demonstrations that are too dangerous to do in front of a class, and a filmed version enables the class to view in complete safety. There are demonstrations that would take far too long to perform during a teaching session and the editing facility makes it possible to condense the important elements into a reasonable time. There are demonstrations that require large and complex props and use the camera to give the audience a particular viewpoint that could not be achieved in a live demonstration. Looping the loop in an aeroplane or a time-lapse film of the train journey from London to Brighton played back in just two minutes are the kind of things I have in mind. Perhaps the most superb example of this kind of film used to make an extremely important teaching point is the film, made for the MIT Physical Sciences Study Committee, called 'Frames of Reference'. It begins with a head and shoulders figure beginning to talk; another figure appears, apparently upside down, and begins an argument as to which of the performers is really upside down. It eventually becomes clear that the first figure was upside down relative to the laboratory reference frame but the camera was also upside down. There are many more demonstrations involving fixed and moving cameras and other tricks, and I can think of no other method by which the idea of relative motion and reference frames could be introduced in anything like such an effective way. The film is in black and white and now some twenty years old, but it is still enormously valuable in teaching undergraduate physicists.

Another important use of film is for demonstrations that only work *sometimes* and, while I do not regard this as an excuse for not trying, it is useful to have a back-up record of a successful performance to use in case the live attempt fails. One classic group of experiments that fall into this category is that of electrostatics.

There are dozens of experiments in the electrostatics category that are spectacular when they work, but rather dismal if they do not. I remember hearing of an incident in which a lecturer was due to talk to a large group of firemen about the part played by electrostatic sparks in starting fires and explosions. He decided to start with a dramatic demonstration; he was to walk into the room without a word and to peel off a woollen jumper, worn over a silk shirt, and to apply his finger to a gas jet, thereby producing a spark and igniting the jet. It worked successfully every time he tried it at rehearsal. But at the actual lecture nothing happened and so his lecture started with a rather limp explanation of what should have happened. What went wrong? He had clearly omitted to realise that the absence of several hundred people emitting water vapour at the rehearsal enabled him to perform the experiments in a dry atmosphere, but at the lecture proper the enormous amount of water vapour produced by the audience was enough to ruin any electrostatic demonstration. A film of the successful rehearsal would at least have partially restored the loss of face!

There are, of course, many experiments that can only be done in a research laboratory and need so much equipment that they would be virtually impossible to set up in front of an audience. Some of the marvellous films on the quite extraordinary properties of liquid helium, made by Professor J F Allen, are clearly in this category.

2.5.3 Use of complete programmes

So far in this section I have implied the use of film to illustrate a specific point. But what about the use of complete programmes? I would always press for live lecture demonstrations wherever possible, but a filmed lecture is certainly better than nothing. If the film is of a presentation in lecture format then, clearly, the lecturer is deciding what should go in and how the ideas should be put across. But if the presentation is in the format of a television programme with a mix of studio demonstrations, outdoor sequences, film clips, etc, together with a spoken commentary, the question of who determines the content and who takes part becomes more controversial.

I suppose that some of the most influential science education programmes have been natural history films, such as Sir David Attenborough's 'Life on Earth'. Obviously a live lecture is impossible when the material comes from a wide range of locations and, in any case, as he pointed out himself in the 1987 Richard Dimbleby Lecture, the cameraman can wait days for a particular shot lasting only a few seconds; even if the creature concerned could be brought into a theatre the chance of any particular bit of behaviour occurring at the key point in a lecture is almost zero, and even if it did, the audience would probably not be looking in the right direction at the crucial moment! Clearly film, though time-consuming and costly, is the only solution—and once recorded an event can be seen many times.

What puzzles me is why the television programme makers adopt such a totally different approach in programmes about the physical sciences. There are many physical phenomena that are every bit as photogenic as natural history. The production of a spectrum, the marvellously coloured patterns produced by crystals in polarising light, the appearance of all kinds of surfaces under the electron microscope and countless more make a superb visual impact. Yet in most programmes on physical sciences produced by the networks a presenter who is not a scientist tries to explain and usually succeeds in conveying the impression that 'it's really too difficult for people like you to understand'. However, I should not let one of my pet hobby-horses intrude or I shall be in danger of straying wildly off my subject.

2.5.4 Variable-speed photography

Of course one film possibility that can never be matched by live demonstration is variable speed; slow-motion studies of rapidly moving objects or time-lapse photography to speed up very slow changes provide powerful educational material in many different fields.

I said at the beginning of this section that I would use the word film to embrace both film and video, and most of what I have said so far applies equally well to both. But there are significant differences and I should mention these before closing the section. Until relatively recently film was the cheaper of the two and had the advantage that editing was possible even by amateurs. The position is now very different; colour video cameras and associated recording equipment have dropped in price and editing is no longer such a problem, and at the same time the cost of film stock has risen very sharply. The new portable video cameras that incorporate the recorder and are fitted with a microphone have now come down below £1000 and produce remarkably high-quality results, even in the hands of amateurs, and offer the very real possibility of teachers purchasing the equipment. Developments in video techniques are following each other very rapidly and it is now possible to 'drive' a video camera with a microcomputer to permit time-lapse sequences to be produced. An excellent example of this is described by Graham Sumner[51] as applied to the recording of cloud sequences over long periods of time and relating them to meteorological data.

2.6 CLOSED-CIRCUIT TELEVISION

2.6.1 Introduction

By closed-circuit television I mean the use of a camera, which may or may not be in the lecture room, which is relaying 'live' pictures to monitors in the lecture room without intermediate recording. Use of a camera in another room has been made, for example, in relaying pictures of a surgical operation in progress into a lecture theatre and this has some obvious benefits. I have tried using a remote camera together with an audio link to enable a class to see what is going on in a research laboratory. So far, however, this has not been very successful and tends to be rather slow; getting just the right picture and resetting the camera between shots takes quite a time and the audience can quickly become bored unless the whole thing is very carefully planned beforehand. Then some of the immediacy is lost and, on the whole, I have found it better to take a longer time and make a video recording that can be edited down to the parts that really matter.

However, I find the use of a camera in the lecture room to augment demonstrations invaluable. It can also be a very cheap addition to the lecturer's equipment. I use a very simple black and white camera of the type used for security surveillance, though I find a zoom lens a useful addition, and an ordinary television set for displaying the pictures. The camera has no viewfinder, but the difference in price between this camera and more sophisticated ones with finders pays several times over for a small portable television that I have in front of me on the bench to fulfil the functions of a viewfinder and to enable focusing to be done.

2.6.2 Displaying small details

I use such a system, in effect, to enable the whole audience to see exactly what I am seeing when I stand close to the equipment. And if the camera is fitted with distance rings between the lens and the camera body it is possible to move right in to produce magnified images of parts of the apparatus. I shall give two examples of the use of this system, one in optics and the other in sound.

Demonstration 2.10

It is extraordinary how few people have seen diffraction phenomena, except perhaps the haloes produced by a steamed-up window when a distant street light is viewed through it, or the two-slit pattern (Young's fringes). (I should perhaps point out here that, in my terminology, diffraction occurs when a wave interacts with an object, e.g. at the double slit, and the resulting scattered waves interfere to produce a pattern on the screen.) But it is very easy to demonstrate a whole range of diffraction phenomena if a simple closed-circuit television is available. Because a great many of my lectures are away from my base I have devised the lightweight, portable substitute for an optical bench that was described in Demonstration 2.3, p. 65, and is illustrated again in figures 2.10 and 2.11. The objects, as before, are photographic reductions on to glass slides with an overall diameter of about 3 mm. A glass screen with a finely ground surface is placed about 1–1.5 m on the far side of the object and, without any further lenses, a diffraction pattern of sorts can be seen on it.

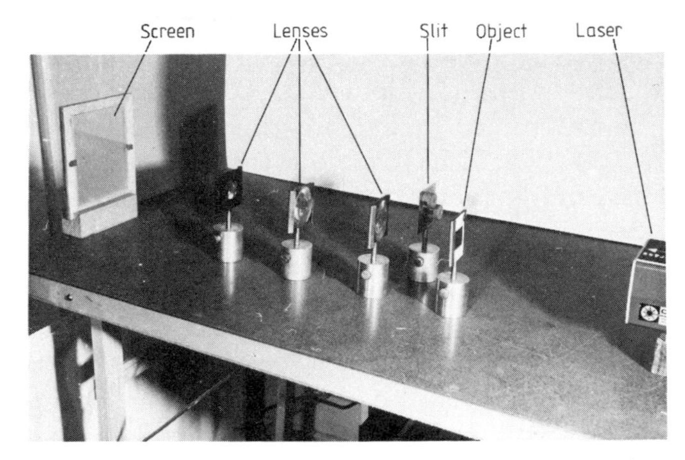

Figure 2.10 Portable diffractometer; the razor blade and magnet arrangement for producing a variable slit is set off line and all the components are placed much closer together than they would be in practice.

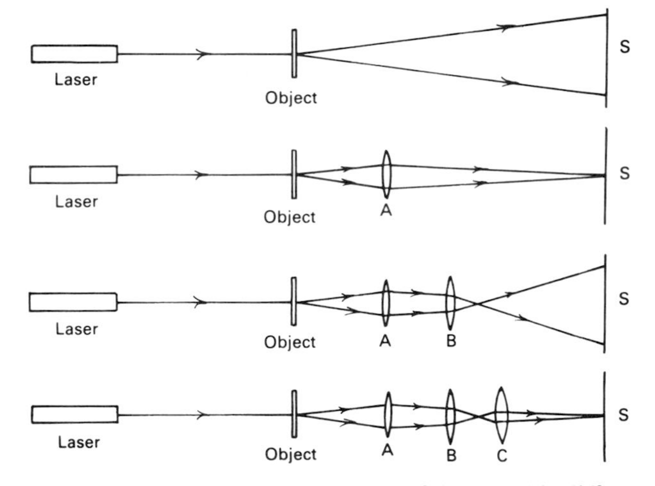

Figure 2.11 Diagram of the arrangement of the portable diffractometer to show the effect of the successive addition of lenses. (From Taylor C A 1985 *Proc. R. Inst.* 57 234.)

The black and white television camera, with the appropriate distance rings between the lens and the body, is focused on the ground glass screen and a whole audience can see the resulting pattern. Some people advocate removing the lens of the camera and letting the pattern fall directly on to the sensitive surface of the camera tube. I have two points against this. First you have less control of size; with my arrangement the image size is controlled by the distance between camera and screen, or by adjustment of a zoom lens. But even more importantly, if by any chance the object is removed and the beam falls on the tube the sensitive surface may be damaged in seconds.

If an auxillary lens is introduced as at A in figure 2.11, an image of the object can be produced; a second lens, B, can be positioned to produce the diffraction pattern again on the screen; this is a much clearer version than that produced with no lenses, as the spot forming the basis of the pattern is now a small focused point rather than the unfocused beam of the laser. A third lens, C, brings back the image of the object again, and so one can go on demonstrating the alternating relationship of object and image (roughly that of a Fourier transform and its inverse). Reverting to the arrangement of figure 2.11, with lens A alone, if it is moved back and forth along the axis of the system with a very regular object (e.g. fine gauze or fabric) a whole series of so-called Fourier images can be shown. How can one distinguish the 'correct' image? The simplest way is to introduce a small defect, e.g. a speck of dirt, on the object blocking up one or two holes. Only one of the images will reproduce the defect; all the others will average it out (figure 2.12).

This system can also be used to demonstrate image processing. In the arrangement with all three lenses, a slit is introduced to cut down the diffraction pattern in a desired way and the modified image is seen on the screen; by removing lens C the portion of the diffraction pattern being recombined to form the image can be shown. Figure 2.13 shows an example of the use of this technique, using a rather more professional optical system, in which one can see not only that the reduction of the aperture

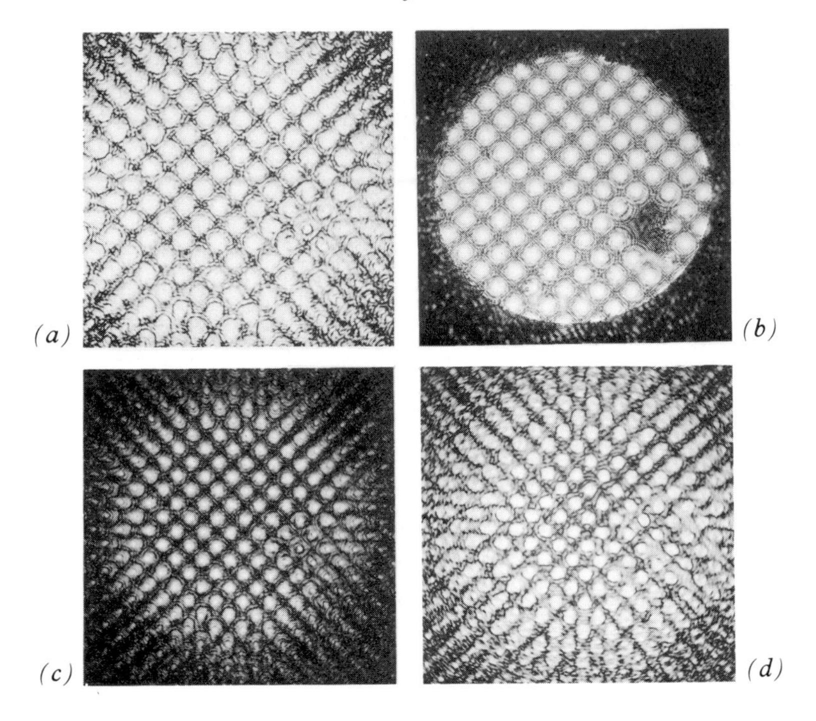

Figure 2.12 Four different images of a piece of perforated foil, using the system of figure 2.11 with lens A only; the distance between the lens and the object was adjusted between each picture. (*b*) is the 'correct' one since it is the only one that clearly shows the defect, the boundary and the correct number of spots. (From Taylor C A 1985 *Proc. R. Inst.* **57** 235.)

raises the instrumental limit on the resolution, but also that false detail may be introduced (e.g. the additional small spots in the central group in 2.13(d) and the fact that in 2.13(f), the central group appears to be nine blurred spots, whereas the object has 25 spots).

Demonstration 2.11

One of the most useful demonstrations in a lecture on sound is the display of waveforms on an oscilloscope. Large-screen demonstration oscilloscopes do exist, and I have used them from

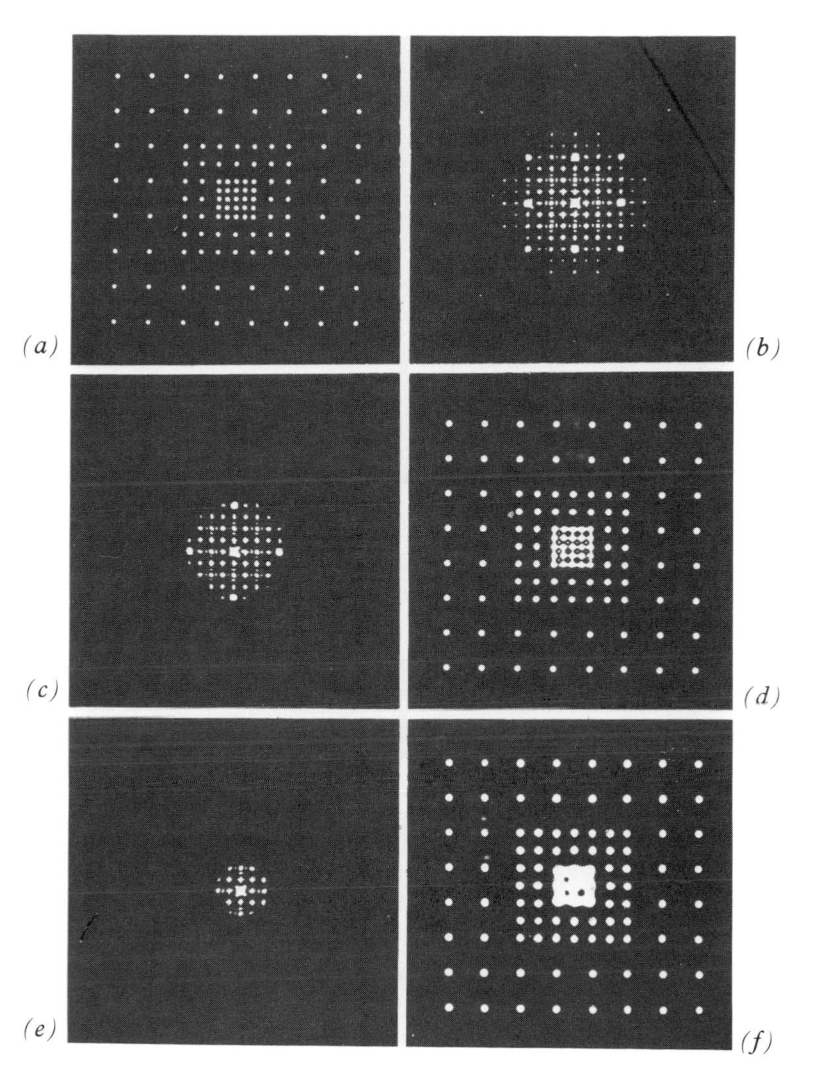

Figure 2.13 Illustration of the effect of the reduction in resolution for a regular object. (*a*) is the object; (*b*) is its Fraunhöfer diffraction pattern; (*c*) is (*b*) restricted by an aperture; (*d*) is the image recombined from (*c*); (*e*) is as (*c*) but with a smaller aperture; (*f*) is the image recombined from (*e*). (From Taylor C A 1987 *Diffraction* (Bristol: Adam Hilger) p. 38, figure 4.4.)

time to time, but they usually have a rather restricted range of controls. I find it much more effective to use a laboratory oscilloscope with a 100 mm wide screen with the same surveillance camera as described in Demonstration 2.10 above to relay the image on to large monitors. Again I use either a fixed-focus lens and distance tubes, or a zoom lens. One of the main advantages of this scheme is that I can take my own oscilloscope and camera and ask the organisers of the lecture to provide the monitors. In this way I make sure that the devices that need constant adjustment, i.e. the oscilloscope and camera, are ones with which I am thoroughly familiar, but at the same time I do not have to carry very large items of equipment.

The same camera can also be used for displaying other demonstrations to the audience; for example the sand figures produced by bowing a Chladni plate (see Demonstration 3.6, p. 122), or the behaviour of a singing flame as the tube comes in to resonance (see Demonstration 1.15, p. 31).

2.7 MICROCOMPUTERS AND INTERACTIVE VIDEO-DISC

2.7.1 Microcomputers

In an article entitled 'Computers and the future of education'[52], Tom Stonier asserts that

> The computer without a doubt is the most powerful pedagogic tool invented since the development of human speech.

and then goes on to list nine reasons in support of his claim. His first, and most important, reason is that the computer is interactive—and the user's response determines what happens next. This kind of use does not really apply to lecture demonstration, except of course in lectures about the science of computing. But later in his list he says

> Computers can explain concepts in a more interesting and understandable manner by means of animated material. No amount of talking, writing, or providing diagrams, can compare with making things come alive on the screen.

And, of course, this applies just as well to a large screen in front of a large audience as to the small screen of a personal computer and hence comes into the category of demonstration in my terms.

But computers are limited by the software available and there are all too few packages available that can be used in this way. And I believe that it may not always be desirable to use an animated diagram, produced by the computer, in place of a real piece of apparatus. The circumstances in which the computer has the advantage are the following.

(1) When it is desirable to demonstrate a process or phenomenon that would be hazardous if demonstrated with the actual apparatus. In this category might come demonstrations of the decay behaviour of radioactive materials.

(2) When it is desired to speed up a process artificially. Again radioactive delay is a typical example where the time scale would in reality be far too long.

(3) Where statistics of large numbers are involved and to do a demonstration on the real-life scale would be out of the question. In this category might be genetic experiments on breeding through many generations which can be telescoped into a short time and deal with enormous numbers if a suitable computer model is available.

(4) Where it is desired to demonstrate experiences with a very large-scale system that exists outside the lecture room and for which a scale model would not be practicable. An example in this category might be weather systems.

(5) Where a demonstration of some kind with real apparatus is to be performed on the bench and it is desirable to record a sequence of results and to display them in graphical or some other diagrammatic form as the experiment proceeds.

2.7.2 *Interactive video-disc*

The main disadvantage of videotape recordings is that it can take quite a time to find exactly the sequence you wish to use out of a two- or three-hour tape (though the technology is improving all the time and some recent tape players perform this function much more rapidly than earlier ones). Yet it is really too costly to use a separate tape for each five- or ten-minute sequence. The video-disc certainly looks promising, though there is still a considerable amount of development to be done. Its advantages are that the picture quality is higher, the discs are extremely durable and, since they are 'read' by a laser beam with no physical contact between the disc and the reading mechanism there is no wear problem. Up to 50 000 still pictures can be recorded on a single disc and any one of them can be found within five seconds. Such a resource has obvious potential illustrating lectures and of course it could equally well be used to store a large number of short animated sequences which again could be located very quickly.

At the present stage of development the main drawbacks are the cost of the reproduction apparatus and the fact that it is not possible to record directly on to disc in order to make one's own programmes. However, the pace of development in this field is so rapid that it could well be that by the time this book is actually in print home recorders may be on the horizon.

It is interesting in passing to note that the home entertainment market has had enormous effects on the costs of electronic equipment of various kinds. Digital, multi-track recorders produced primarily for the music recording industry are now being used in many branches of science and engineering as multi-channel data recorders. Laser diodes, which were developed specifically to read compact discs in home disc players, have now found applications in many areas of science such as medicine, robotics, etc. But the key point is that it is the enormous demand of the home entertainment market that has enabled the costs of development to be absorbed much more quickly than would otherwise have been possible and it is to be hoped the same principle will apply to video-disc recording.

Part 3

The Practice of the Art

3.1 GETTING STARTED

The potential barrier that has to be surmounted by anyone deciding to venture upon lecture demonstration, whether it be in a school classroom or for a large-scale public lecture, is considerable. But I hope that I have said enough already to convince you that the rewards are also considerable. Contemplating a lecture that is to be illustrated with demonstrations is a bit like an artist contemplating a blank canvas or a writer facing the notorious blank page. Everyone has a different approach and to some extent the approach will vary with the purpose for which the lecture is being given. If the purpose is to teach about a particular topic, then the teaching material must be settled first and the appropriate demonstrations slotted in. But if the lecture is in the nature of a 'popular' presentation then my experience has been that it is best to collect together the appropriate number of demonstrations and then to write the lecture round the demonstrations. In this way a fully integrated presentation can be produced. It seems to me that, whatever the purpose, integration of the demonstrations into the lecture material is vital. Some lecturers give a straight lecture and then, almost as an aside, give a few relevant demonstrations at the end.

And there are some famous examples of university lecturers who would go through the whole term just using the blackboard for illustrations and then give a splendid end-of-term performance of demonstrations relating to the whole term's work. Of course I can see the arguments in favour of this, not least that it reduces the amount of work to be done by both lecturer and technician in preparation. But, to my way of thinking the result is second best. There is no better time to do a demonstration than when a particular concept or principle is introduced. If the idea, the theory behind it, and the demonstration march together, the likelihood of its making a real impact and being properly understood and remembered is greatly enhanced.

A little earlier in the section I mentioned 'the appropriate number of demonstrations'. How can we arrive at this number? Clearly it depends very much on the audience and the primary purpose of the lecture. In preparing to give the 1971 series of Royal Institution Christmas Lectures I adopted the principle that something in the nature of a demonstration, or a visual or audible happening (i.e. a slide or a recording), should occur at least once every two minutes. For six hours of lectures that meant 180 events and, to allow for the failures and misadventures during the long period of preparation and for those that, as the lecture was built up, did not fit easily into the scheme, this meant finding about 300 possible events. The first hundred or so came quite easily but after that development became rather slower!

It is interesting to note that John Tyndall seems to have used this sort of figure. For example, in a Friday Evening Discourse in 1866 he performed about thirty experiments in the hour on a single theme.

3.2 THE IMPORTANCE OF PREPARATION

3.2.1 Introduction

I cannot overemphasise the importance of careful preparation in a demonstration lecture. There is a story told of a visitor to the Royal Institution in the middle of the last century who was watching preparations for a discourse from outside the open doors of the theatre. He saw a tall lean figure standing at the bench, placing his hand on the bench and leaping on to the bench and over to the floor on the audience side. This feat was repeated several times to the amazement of the visitor who, when he attended the evening discourse was surprised to see that this same lean figure was the lecturer and that in the middle of the discourse when some mishap occurred in front of the bench he calmly leapt across to make an adjustment, using the technique developed during his careful preparation. The lecturer was John Tyndall and this incident typifies the enormous care he took with all aspects of his lectures.

3.2.2 Mental and practical preparation

With regard to mental preparation Sir Lawrence Bragg[53] writes

> A good lecture is a *tour de force*; a good lecturer should be keyed up to a high pitch of nervous tension before it and limp and exhausted after it. In my own experience, the occasions when I have felt confident have been disasters; the confidence comes from a doping of healthy criticism.

While agreeing entirely with this, I am sure that another factor in developing confidence is care in the preparation of both what you propose to say and what you propose to do. As I think I have said elsewhere, I cannot work from a complete script, though I am aware that many people prefer to do so. But whether you work from a script or from notes it is essential to go over in your mind not just the words and the actions, but also the aims of the lecture, what you hope the audience will have gained by the end, and perhaps to think about what questions *you* might ask afterwards in the hope that you can be at least partially prepared for the questions that the audience may produce. Of course there is always the unexpected tricky question, and in the last resort it is far better to give an honest 'Don't know' and a promise to find out than to waffle.

But what sort of practical preparations should be made? I think that there are three sorts of preparation that can be identified. First there is the detailed working out of the best way to do each of the separate experiments that will build up into the whole lecture. Then there is the rehearsing or practising the actual performance of the experiments in a lecture sequence and making sure that the lecture flows smoothly from one to the other. Finally there are all sorts of mundane practical details about the lecture room and its equipment.

3.2.3 The mundane preparations

I will start with these mundane elements, neglect of which I have nevertheless seen wreck the start of a lecture. The location of the switches and fuses for the lights and the power supplies should be known to the lecturer, or to a technician who is to be present. If a technician is to help, it is obviously important to be clear about the division of duties. He or she needs to have a clear idea of who is to change the slides, who is to dim the lights and so on. It is also important to be prepared for disasters; spare fuses and bulbs for projectors and similar apparatus, together with the appropriate screwdrivers should be available. But how will you proceed if there is a more difficult breakdown to correct? Have you an emergency plan that can allow the lecture to continue without slides, or without a hi-fi system? These points are further elaborated in §3.7 (Coping with disasters).

3.2.4 Rehearsal—the complete presentation

Taking the kinds of preparation in reverse order the next to consider is rehearsal. One of the tricks that I have learned from bitter experience is that one must *never* assume that because an experiment worked a few weeks ago when you tried it out that it will necessarily work today. As far as humanly possible I like to try everything out close to the start of the lecture. I even run my slides through the projector, partly to check that they are the right way up and in the right order, but also to check that the particular features that I wish to make clear to the audience really can be seen from all parts of the theatre. It is better to miss out a slide than to use one that cannot be seen properly.

No.	Point to be made.	Demonstration.	Slide.	Slide Number.
1.	Intro. Colour taken for granted.			
	Painters have great sense of colour		Titian	C'137
	Flower gardens spectacular source of colour		Garden	C' 19
2.	But what if we could not see colour?		Garden B&W	C' 18
			Feather B&W	C' 28
			Feather Col.	C' 29
3.	What use is colour? Makes world attractive....not likely to evolve for this. Protection? Food? etc.		Berries B&W	C' 7
			Berries Col.	C' 8
	Useful in Modern World		Cables B&W	C' 30
			Cables Col.	C' 31
4.	Vast Field........select three areas. (a) Physics (b) Colour Vision (c) Colour Perception			
5.	Start with physics Not aware of wavelengths...... Different degrees of refrangibility	Spectrum...Newton's experiment with prism		
	Was aware of rainbows		Rainbow	C 17
				C 18
	Contemporary view was that prism added colour.			
6.	How many Colours?	Quote from Opticks		
7.	Current ideas.....		E.M.Spectrum	C' 11
				C' 9
8.	How could we explain colour vision? Infinity of detectors.	Coloured fabrics approx.in rainbow colours		
	But what about non-spectral purples & browns?	and in non-spectral colours		

Figure 3.1 Extract from a typical running order for a demonstration lecture on 'Colour'.

Obviously the apparatus to be used needs to be available on the bench or on trolleys that can be wheeled in at the right time. And even this sort of operation needs planning. For example if several trolleys are to be used their location when the lecture starts is important; nothing is more irritating than to find that several trolleys have to be moved around in order that the next one needed can be got into position. The audience's attention can be lost in a remarkably short time if some muddle of this kind occurs. If a significant number of demonstrations is involved then the only satisfactory way to avoid problems is to treat the lecture like a theatrical performance with the location of the 'props' at each stage indicated on some kind of script. I cannot work from a detailed script; I find that if I try to write out exactly what I want to say or do the result is a very stilted and unnatural performance. But I do use a 'running order' which helps me to keep to the planned sequence of experiments, helps me to have some idea of the timing, helps me to make quick decisions if something is to be left out because of the time, and, of course, if any assistants are involved it gives them some idea of what is to happen when. Figure 3.1 is a page from a typical running order.

Apart from the kind of checks that I have already mentioned, practice can be very important. A particular demonstration done on its own when there is no pressure of time or no audience present may appear to be very easy to do. But, in the middle of a lecture, when the adrenalin is flowing, when you are trying to think of what you are to say next and why the last experiment did not work quite as well as you had hoped, your fingers may suddenly all feel like thumbs and morale can fall very quickly. But if you have rehearsed the experiment several times the resulting confidence can get you back on the rails very quickly. I say 'rehearsed the experiment *several times*' deliberately because it is surprising how often the element of beginner's luck can cloud the picture. On many occasions in my earlier days I have tried an experiment once and found that it worked well, only to discover that in the lecture proper it did not work, usually because I did something slightly different. But if during rehearsal you can do it several times successfully, the chances are much greater that it will work when the time comes.

One of the problems that I find often occurs is that if a certain piece of equipment is needed several times during a lecture it may not be in the right place after the first time. If the item is cheap and plentiful then it is well worth duplicating; for example in one of my lectures I use a pair of scissors to cut drinking straws to make simple musical instruments in two or three ways and it is astonishing how often, after the first use, the scissors seem to hide themselves behind or under some other piece of apparatus and can't be found on the second occasion! The two possible solutions are either to rehearse putting the scissors down in a particular place, or to have a second pair already in a known position. I recommend both on the principle of belt and braces.

It is well worth giving thought to the best layout of equipment on the bench in order to improve the continuity of the performance and it may be necessary to make auxiliary equipment to help keep things in order. For example in a lecture on colour, using three projectors, there are about fifteen slides that are used several times each and, since darkness is essential, I was always misplacing the slides on the bench and having to break the continuity and put the lights on in order to hunt for them. I now have a simple rack and have trained myself to put slides back in the right position, after which they can always be found by feel. Figure 3.2(*a*) is a picture of this rack and the demonstration is described in §3.6 (How demonstrations evolve), Demonstration 3.7, p. 125.

When I am lecturing on the relationships between science and music I use quite a large collection of recordings that has been built up over a long period. In the early years I used a reel-to-reel recorder; all the illustrations were on one tape and I relied on the tape counter and a numbered list of items to find the right illustration. As the number increased and I used different illustrations, sometimes in a different order, for different audiences, the time taken in finding a given item became undesirably long. I therefore changed over to cassettes. In itself this would not necessarily have improved matters, but cassettes are relatively inexpensive and I now keep several copies of each tape and, during the preparation period before the lecture, I set each cassette at the point where a particular illustration starts and then each time I need an illustration it is an easy matter to slip in the appropriate cassette. Again a simple rack solves the problem of keeping them in order on the bench as shown in figure 3.2(*b*).

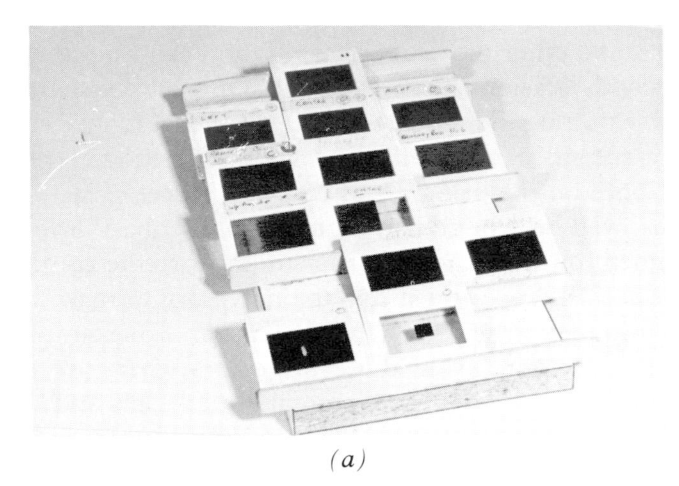

(*a*)

(*b*)

Figure 3.2 (*a*) Rack for slides used in colour demonstrations. (*b*) Rack for cassette recordings used in music demonstrations.

Some of these preparatory ideas may seem a little over fussy, but experience shows that each one gains a little time, improves the continuity, prevents the audience being distracted and losing attention and helps the calmness of the lecturer enormously.

As always, Michael Faraday[54] has some excellent advice:

> ... the whole should be arranged as to keep one operation from interfering with another. If the lecture table appears crowded, if the lecturer (hid by his apparatus) is invisible, if things appear crooked, or aside, or unequal, or if some are out of sight without particular reason, the lecturer is considered (and with reason) as an awkward contriver and a bungler.

3.2.5 Rehearsal—individual experiments

Finally we should consider the development of each individual experiment. This is usually an iterative process and often a lecture has to be given many times before precisely the right way to perform a particular demonstration can be settled. John Tyndall had endless patience in searching for the best way to perform a demonstration. For instance, in an essay on John Tyndall as a lecturer[55] I quoted from his manuscript lecture notes on an experiment designed to show that the musical vibrations that could be produced by striking an iron poker with a hammer could be used to set up transverse standing waves on a piece of 'elastic' tied to it. He began by suspending the poker on a length of elastic from the ceiling and tried many variations of tension, length, etc. He also suspended tuning forks and bells without success. The notes then run as follows:

> 'Thinner elastic—better. Subdivision of string obtained with a brass rod giving a loud though low note. Experiment not very certain, not pronounced; a better method is as follows:-' He then describes a complete change, with the length of elastic placed horizontally so that the tension could be more easily varied: he substituted a piece of dark thread for the elastic and then notes:
>
> 'There is an improvement when seen close at hand, not so well seen at a distance'. Then he tries two forks, one at each

> end; he switches to silk; 'Scarlet silk gives a beautiful gossamer appearance when vibrating before a black background. White silk seen best at a distance'. And so it goes on over a considerable period of time. The first entry is dated 1 May 1866 and the last 30 May 1866.

As can be seen, Tyndall was very preoccupied with the visibility of the experiment to the audience. He passionately desired his audience to see and to understand the physics as clearly as he did himself, and no trouble was too great for him in achieving this end.

3.3 THE IMPORTANCE OF VISIBILITY

3.3.1 Introduction

Nothing is more frustrating to an audience than to be unable to see what is going on in an experiment. Generally speaking the larger the apparatus, the more effective the demonstration is likely to be (see reference to Lichtenberg, p. 20), though of course there are usually practical considerations that need to be taken into account and set an upper limit to the size of the apparatus.

3.3.2 Slides

In §3.2.4 I mentioned how important it is to make sure that the audience can see the contents of slides clearly. I have attended many a conference lecture by an inexperienced lecturer whose slides were so full of tiny, pale text that it was quite impossible to gain any information from them. Furthermore I have seen a lecturer leave a slide on the screen for only a few seconds and then carry on to the next stage assuming that everyone has grasped the points made by the slide!

There is a very simple test that all lecturers should be aware of; hold the slide about 30 cm from your eyes and if you cannot read the lettering clearly your audience will not see it when it is projected. Of course this assumes that the focal length of the projection lens and distance away of the screen are all suited to the size of the lecture room. Charts and posters should all conform to the same sort of rule—that is that they must be capable of being viewed comfortably from a distance equal to about six times the width (this agrees with the 30 cm viewing distance for a 5 cm wide slide). Kenny[56] discusses in much more detail the choice of type size and so on to be used in preparing visual material and I recommend any budding lecturer to take this advice very seriously. Incidentally, it is just as annoying to have a slide on view long after it has fulfilled its purpose as to show it for too short a time. If, therefore, you are using an automatic projector which cannot be switched off from the bench, it is important to provide blank slides in between each sequence.

3.3.3 Size of apparatus

But, of course, the need for visibility applies equally well to demonstration experiments. Even if the demonstration consists simply in showing the audience some piece of equipment it is important that they should be able to see clearly what the essential features are. It seems almost impertinent to mention the first and most obvious point—but I have often been present at lectures when it has been ignored. It is that the object should be held still for a long enough time for the audience to take in the details. On several occasions I recall seeing a lecturer pick up an object from its original position out of sight of the audience, wave it in the air for a few seconds and return it to its hiding place! If the object is too small to be seen clearly there are several alternatives open: a photograph, or a drawing can be displayed with a slide or overhead projector; a closed-circuit television camera can be used—though here again it must be stressed that it is important to hold the object quite still for the appropriate time; and the third possibility is to use a large scale model. The last suggestion involves a good deal of work but is often the most satisfactory solution, especially if the model can be taken apart to show the construction, or moved to show the mode of operation.

In *Advice to Lecturers*, from which I have already quoted many times, Sir Lawrence Bragg writes[57]

> Faraday had much to say about experiments that was very wise. The best experiments are simple and on a large scale, and their workings are obvious to the audience. The worst experiment is the one in which something happens inside a box, and the audience is told that if a pointer moves the lecturer has very cleverly produced a marvellous effect. Audiences love simple experiments and, strangely enough, it is often the advanced scientist who is most delighted by them.

There are, of course, constraints on what one can do, especially if the lecture is to be delivered at a venue that involves transporting the apparatus some distance, but in general my advice is to make the apparatus as large as is practicable. I have already referred to good demonstration as, in some ways, a theatrical performance and large scale apparatus adds a degree of theatricality that is well worth while. For example some years ago, on the first occasion on which I delivered a demonstration lecture on 'Colour', I wanted to draw the audience's attention to the misguided way in which schools (and indeed, I regret to say, art colleges, as is borne out by very recent publications) teach the basic theory of colour mixing. I decided to use dyes that could be mixed in various ways, and also, since I was fortunate to be able to present the lecture on a full-size theatrical stage, that I should use very large quantities. The demonstration proceeded as follows:

Demonstration 3.1

Five very large photographic developing dishes, each about 600 × 400 mm, were divided down the middle lengthwise cementing in a plastic partition. If filled with liquid, each side would hold about 2 l when filled to a depth of about 20 mm.

Fifteen two-litre buckets were then prepared containing the following proportions of photographic dyes diluted so that when 2 l is poured into a tray the colour can be seen vividly by the audience. I will not give quantities here because this will depend on the actual dyes used. Three basic dyes are used—magenta (M), yellow (Y) and cyan (C)—and, they should be chosen as nearly as possible so that they absorb only one of the three additive primaries, respectively green (G), blue (B) and red (R). Table 3.1 shows how the 15 buckets are prepared, table 3.2 how they are mixed, and table 3.3 the result.

Table 3.1 The contents of the 15 buckets in litres.

1	1 M + 1 Y = 2 R	6	1 Y	11	1 C
2	0.5 M + 0.5 Y = 1 R	7	2 Y	12	1 M
3	0.5 M + 0.5 C = 1 B	8	1 Y	13	2 M
4	1 M + 1 C = 2 B	9	1 C	14	1 M
5	0.5 M + 0.5 C = 1 B	10	2 C	15	1 Y

Table 3.2 The mixing sequence in terms of bucket numbers.

Tray number	1	2	3	4	5	6	7	8	9	10
Stage 1	1	–	4	–	7	–	–	–	–	–
Stage 2	1	2 + 3	4	5 + 6	7	–	–	–	–	–
Stage 3	1	2 + 3	4	5 + 6	7	–	10	–	13	
Stage 4	1	2 + 3	4	5 + 6	7	8 + 9	10	11 + 12	13	14 + 15

Table 3.3 The results of the mixing.

Tray 1	Red
Tray 2	R + B = (−B − G) + (−R − G) = −W − G = dark magenta
Tray 3	Blue
Tray 4	B + Y = (−R − G) + (−B) = −W = black (or more probably dark grey)
Tray 5	Yellow
Tray 6	Y + C = (−B) + (−R) = green
Tray 7	Cyan
Tray 8	C + M = (−R) + (−G) = blue
Tray 9	Magenta
Tray 10	M + Y = (−G) + (−B) = red

This demonstration proved to be very successful for several reasons: first it was easily visible to all the 500 children present, secondly it worked and showed very clearly that the mis-chosen subtractive primaries of red, blue and yellow give very muddy colours when mixed whereas the correct primaries of magenta, yellow and cyan give much more brilliant mixtures, and thirdly there was an element of drama, or even comedy, in sloshing about large buckets full of dye (see figure 3.3).

Figure 3.3 '. . . an element of drama, or even comedy, in sloshing about large buckets full of dye'. (Photo by R S Watkins.)

3.3.4 Visibility of details

At the end of the last section I mentioned John Tyndall's anxiety that his audience should see clearly. I am sure he would have approved the use of closed-circuit television to produce enlarged images of small-scale demonstrations (see §2.6.2) and of lasers to give greatly increased brightness to optical effects. But even though such modern devices were not available to him he showed remarkable ingenuity in increasing visibility. One further development of his demonstrations of standing waves on a string is worth describing.

Demonstration 3.2

The apparatus consists of a more or less standard arrangement for showing standing waves with the usual steel wire replaced by platinum (I use Nichrome or some similar alloy). One end of the wire is attached to a tuning fork and the other passes over a copper bridge to a peg with which the tension can be varied. A current is passed between the fork and the copper bridge so that the wire begins to glow. The tension can be adjusted to give different numbers of nodes along the wire. The wire is cooled by its vibration in the air but, of course, at a nodal pont where there is virtually no movement the cooling is much less and these points glow more brightly than the loops between the nodes. As the wire settles into a steady state of vibration the nodal points begin to shine out even more brightly; this is because the resistance increases with temperature and so more heat is generated at the nodes than in the loops, so the process is augmented.

When performed in a darkened room this is a most beautiful and graphic demonstration. I sometimes use a pulley and weight to give constant tension and then drive the wire with an electromagnetic vibrator and variable frequency oscillator; provided that the oscillator has a fine frequency control it is possible to adjust the system to give several different modes.

Demonstration 3.3

When a wire or thread is driven at a frequency in between those of two modes some very odd, constantly changing patterns arise and usually the wire vibrates in many directions other than those in the vertical plane contained by the length. Tyndall used to demonstrate what was happening by shining a beam of light from a vertical slit across the thread so that only a millimetre or so length of the thread was illuminated. When vibrating in one of these intermode states, the observer sees this spot of light executing figures, rather reminiscent of Lissajous' figures. The disadvantage of this method is that even though a silk thread can give quite large amplitudes of vibration (several centimetres) it is nevertheless still difficult to see the figures from the usual audience distance. I have used an up-to-date version, which needs only small amplitudes, in which a 1 mW He–Ne laser beam illuminates the wire and a closed-circuit television camera displays the resultant patterns to the audience.

3.4 THE IMPORTANCE OF PRESENTATION

3.4.1 Layout

The actual way in which a demonstration is introduced and performed can have a profound influence on its effectiveness. Barlex and Carré[58] have some wise things to say about making lessons visually appealing; they are discussing teaching in school, but the same principles apply at any level. They show two photographs of alternative layouts for a demonstration lecture on the preparation of chlorine. In the first there is a jumble of reagent bottles behind the lecturer, leftovers from the last lesson still piled on the bench, yesterday's homework still written on the blackboard and the apparatus set out in a rather confusing way so that it is not immediately obvious which tube is going where. In the second photograph there is nothing visible except the apparatus itself and this is arranged very clearly so that students can see exactly what is connected to what. Obviously it is not necessary to elaborate on the important principles illustrated by the two pictures.

I mentioned in the last section the importance of keeping an object in view for a reasonable period of time after it is brought out from a hiding place. I did not intend to suggest that it should not be hidden in the first place; in fact I find that one very useful trick in presentation is to keep certain items hidden away until they are needed. For example, in the musical saw demonstration (§2.3.3), from which several important points about musical instruments can be brought out, the dramatic effect (and the consequent likelihood that the physics points made will be remembered) is enhanced if the saw is hidden under the bench and only produced when the demonstration is to be done. Another related point is that any apparatus for a demonstration that might have to be cut out because of shortage of time is best kept concealed; it can be quite distracting for an audience to be kept puzzling over the purpose of an item of equipment that in the end is never used. However, I have to admit that this is a trap that I have fallen into myself on several occasions.

3.4.2 Presentation technique

Inevitably in a book of this kind there are considerable amounts of overlapping between the various sections and this is particularly true between §2.3 on audience psychology and this section on presentation. In that section I spent some time discussing the way in which the brain can perceive and absorb visual or oral messages more effectively if it has already been primed in some way. Sir Lawrence Bragg makes this point very clearly[59]:

> The wrong way is to do the experiment, ask the audience if they noticed this or that, then explain what this or that meant. The right way is to start by explaining the significance of the effect you are aiming at producing, tell the audience what to look for, and then, after a pause to make sure you have their attention, to bring it off. These tricks are important because they are all part of fixing your message in the minds of the audience; they have the humble but necessary function of the hypo in fixing a photographic exposure.

I am quite convinced that this is one of the important keys to effective presentation of demonstrations and my rules would be something like the following: first describe the apparatus that is to be used and remember that although it is extremely familiar to you even a relatively specialist audience may not immediately recognise your particular version; next describe clearly what you are going to do; and finally explain what the audience may be expected to see or hear.

As always there are occasional exceptions to this rule, but even then it is important that the audience knows exactly what is expected of them. For example when I lecture to audiences of children about colour one of the experiments is on after-images.

Demonstration 3.4

The audience is told that a picture of an object is going to be projected on to the screen and they must stare hard at the centre of it without moving their eyes while we all count up to twenty. I tell them that, after that, something will happen but that they must go on staring at the screen. I then project a picture of an elephant in solid green on a black background, count to twenty with the children, on twenty, replace the slide with a completely clear slide giving a white patch on the screen. There is always a gasp, followed by giggles as they see the magenta after-image appear.

I have tried several different ways of presenting this demonstration but this seems to be the most effective. Of course the explanation must then follow in terms of the tiring of the green receptors in the region of the retina on which the image of the elephant falls and the relative freshness of all three receptors in the region on which the black surroundings were imaged.

3.5 THE PROBLEMS OF THE TRAVELLING LECTURER

3.5.1 The value of a visiting lecturer

When I was in about the second or third year of my primary school we had two happenings that I can still remember very clearly, although my recollections of most things that happened then are extremely vague. Both were visits by a travelling lecturer. One, as I mentioned in §2.4.1, brought slides, and I was very intrigued by the magic lantern and spent most of the visit pondering on how it worked, with the result that I did not take much notice of the lesson material. But the other brought a suitcase full of exciting looking bits of glass apparatus and proceeded to do experiments designed to show what alcohol does to the human body—and I could describe every experiment now, almost sixty years later. But, although this certainly shows *my* response to demonstrations, the significant point here is that both visits in themselves caused great excitement because there was a new face, a new voice and someone who was not going to set tests or give marks.

I am convinced that this sort of visit to schools is of immense value, and, of course, in the later years at school it is possible to invite lecturers to come along and demonstrate effects that are related to the curriculum but perhaps beyond the technical resources of the school. Visiting demonstrators at universities have been used almost since demonstration lectures began, and many bodies world-wide such as the Institute of Physics and other learned and professional institutions in Britain, the National Science and Technology Centre in Australia, the Singapore Science Centre, the Lawrence Hall of Science in the USA and many others, invite lecturers to travel from the other side of the world with their apparatus.

But some particular problems arise and it may be worth recounting some experiences that may be helpful to anyone embarking on a journey with equipment.

In 1872–3 John Tyndall embarked on a lecture tour of the USA which proved immensely popular. In his book *Six Lectures on Light*[60] he wrote:

> . . . last year I was honoured with a request so cordial, signed by five-and-twenty names, so distinguished in science, in literature, and in administrative position, that I at once resolved to respond to it by braving not only the disquieting oscillations of the Atlantic, but the far more disquieting ordeal of appearing in person before the people of the United States. . . .

> ... I was given to understand that a course of lectures, showing the uses of experiment in the cultivation of Natural Knowledge, would materially promote scientific education in this country. And though such lectures involved the selection of weighty and delicate instruments, and their transfer from place to place, I at once resolved to meet the wishes of my friends, as far as time and means at my disposal would allow.

To convey such a vast collection of equipment took a great deal of organisation and it is not too surprising that it was only as a result of persistent invitations that Tyndall responded. At least he did not have to cope with the varying voltages and types of socket that occur around the world; he took his own batteries.

But what about the travelling lecturer today? I shall divide my remarks into four sections, dealing with successively more ambitious demonstrations.

3.5.2 Very portable demonstrations

In my first category are the kind of demonstrations that are very easy to transport, possibly in a pocket or brief-case. Slides or transparencies for the overhead projector come into this category and, although it seems very obvious, and has already been stressed in §3.2.4, I cannot recommend too strongly that the lecturer should arrive early enough to be able to run through the slides, check who will operate the projector, who will control the lights, what signals, if any, can be given to the projectionist, and so on. Nothing is more irritating to an audience than slides in the wrong order or the wrong way round. I remember an auspicious occasion when an English speaker who knew no German was addressing an international audience in a German theatre where the projectionist knew no English. There are eight possible orientations for a slide in a projector and I am sure he tried every one, and some more than once, and by the time the right orientation had been found the lecturer had lost the thread of his lecture and the attention of the audience completely.

But there are other small demonstrations that can easily be carried. One of my favourites uses just a handful of balloons; I use it when I am asked to talk about the nature of physics, possibly at a careers convention or a similar occasion. The discussion starts with the origin of the various colours, goes on to discuss the reasons why rubber can stretch to such an extent, the cooling effect when a stretched balloon is allowed to contract suddenly, the behaviour of the gas inside when a balloon is blown up, the musical (?) sounds that can be produced by allowing air to escape through the double reed formed by stretching the neck and finally the jet propulsion effect when the balloon is released. Although this is so trivial that it can hardly be classed as a demonstration, I have nevertheless been surprised at how many people have reminded me of it on subsequent meetings and it clearly helps to fix the ideas in the mind.

3.5.3 Taking some of the demonstrations on tour

The next category contains those demonstrations for which the lecturer carries with him some specialist items but relies on the local organisers to supply commonly available items of equipment. This arrangement holds particularly for lectures abroad involving air travel, where obviously excess baggage charges need to be avoided. For example, if I am lecturing abroad on 'Physics and music', I take with me various instruments and devices for illustrating the physics of instruments, together with cassettes on which I have recorded aural illustrations of various kinds. But I ask the local people to provide a stereo amplifer, loudspeakers and an oscilloscope to display waveforms. I take my own cassette player because it is vital to have every item properly indexed against the counter on the player and no two players match accurately enough. If, for some reason, I cannot take my own player, then it is vital to arrive with plenty of time to run through my cassettes and re-index them on the player provided. The recently developed professional-quality players of the 'Walkman' type have been a boon because they are so small and one can remind oneself of the contents, and set the tapes at the correct starting points, using headphones in a hotel room or even on a plane.

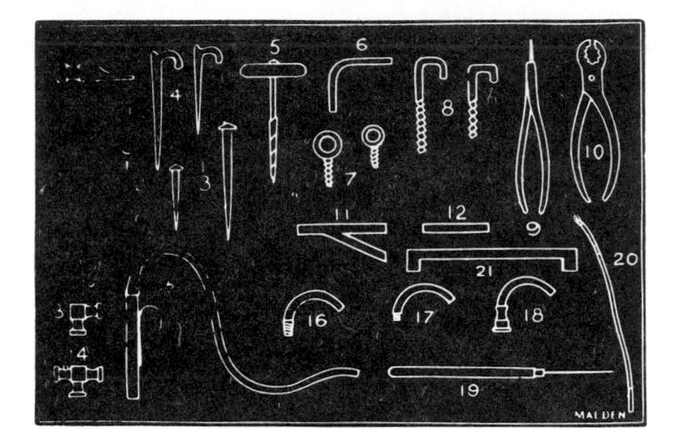

This list will be found a useful reminder to lecturers leaving home as to what they should take with them :—

1, Hammer; 2, Tacks; 3, Nails; 4, Driver Hooks; 5, Gimlet; 6, Bent Rod for turning screw eyes; 7, Screw Eyes; 8, Screw Hooks; 9, Lime Nippers; 10, Gas Pliers; 11, Y Connector; 12, Brass Connecting Piece; 13, Elbow Joint for gas fittings; 14, T Joint for ditto; 15, Connection for hooking to gas burners; 16, Gas Nib with bent tube; 17, Ditto; 18, Ditto; 19, Gas Broach for clearing jets; 20, Wax Taper; 21, A Piece of Wood, varying in *length* with the *width* of the slide-holder, and intended to assist in centring by raising the picture.

Figure 3.4 Lecturer's 'List of sundries' from *The Magic Lantern* by 'A mere Phantom', 1870.

But there are hazards in this kind of arrangement. I have several times arrived to give this kind of lecture and found that our combined resources of plugs and leads do not provide a way of connecting my player to the sound system, or that the audio-visual aids department had been asked to provide the amplifier and assumed that the P.A. system sited up in a control box at the back of the theatre would be alright without realising the problem of connecting my cassette player from the stage to the box. In a book entitled

The Magic Lantern dated 1870[61] there is an interesting diagram (figure 3.4) of a lecturer's 'List of sundries' to be taken. This may be out of date but the principle still holds, and figure 3.5 shows my modern equivalent—a collection of home-made metal boxes (to provide screening) with the various sockets all connected in parallel which enable almost any lead to be connected to any other, together with a soldering iron, solder and screwdrivers, etc.

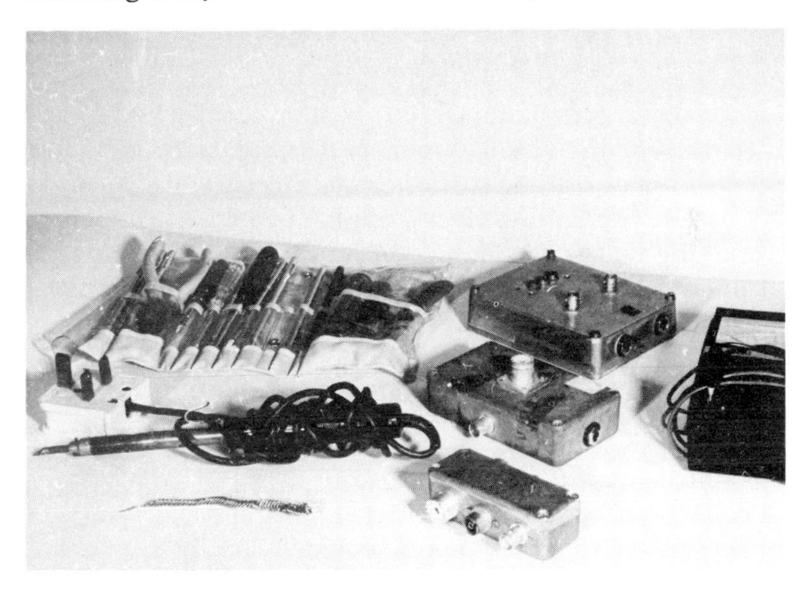

Figure 3.5 Some of the items from my emergency kit.

3.5.4 Relying on local organisers

In the third category are fairly elaborate demonstrations which the local organisers are asked to provide or build to specifications provided by the lecturer. This can work well but there are two provisos that have to be made. First it is essential that the lecturer arrives with plenty of time to adapt, correct or modify the equipment requested and secondly the instructions cannot be too precise. Common names of equipment differ from country to country and descriptions that appear entirely adequate to you, because you have a very clear idea in your mind of what you want, may not be so clear to those reading them. Diagrams help enormously and it is important to explain exactly what the apparatus is intended to do or to illustrate as well as to give the basic constructional or assembly details.

3.5.5 The complete travelling show

The final category is the large travelling lecture that has to be treated more like a circus or theatre company. The Faraday Lecture of the Institution of Electrical Engineers is a typical example. It involves a team of people who transport everything that is needed in vans and build the set at each location. I mention it for completeness, though demonstration on this scale is almost outside the scope of this book. I have on one or two occasions undertaken lectures almost on this scale and the three points that seem to me to be the most important are:

(1) visit the theatre or hall before hand, make measurements where appropriate, check availability of benches, tables, etc, if they are to be provided locally and check visibility of table tops, etc, from various parts of the auditorium;

(2) try out all the experiments at your base as near to the lecture date as can be arranged;

(3) allow π times as long for setting up at the location as it takes on your home ground!

3.6 HOW DEMONSTRATIONS EVOLVE

3.6.1 Introduction

It is rare to find that a new demonstration is completely satisfactory on the first occasion of its use. Occasionally all that is needed is more practice in presentation. Sometimes a different style of presentation is needed to make the desired impact and often in the middle of a lecture I have suddenly thought of an addition or modification that results in a great improvement. Sometimes a demonstration evolves gradually over a period of some years, partly as a result of accidental happenings during the lecture, or perhaps as a result of comments by members of the audience or by other lecturers. And sometimes the evolution is forced in the first instance by the need to incorporate a demonstration in a travelling lecture after introducing it first on one's home ground where weight and size is of less consequence.

3.6.2 Practice in presentation

An example of the first category (more practice) would be my use of a rubber cord to demonstrate modes of vibration.

Demonstration 3.5

There have been many suggestions for the best kind of cord with which to demonstrate standing waves. Some people use gas tubing, some ordinary clothes line, I have even met the use of rubber tubing filled with mercury! But I have no doubt that the best material is 5 mm diameter soft white rubber cord of the type that used to be used for vacuum seals. A length of about 5 m seems to be the most useful. One end is held firmly by an assistant, and it is important that the assistant does not attempt to move in sympathy with the demonstrator at the other end. I send a wave along the rope by plucking it as though it is a larger-scale model of a guitar string. This has two purposes; the explicit one of showing the audience that when a string is plucked a wave actually travels up and down, and a more covert purpose which is to give me an idea of the fundamental frequency of the cord and to adjust the length and tension to get this in the right range. Then, having got the fundamental frequency in mind, I can oscillate my end of the rope up and down at that frequency and hence produce the fundamental mode with a node at each end without too much trouble. But it took quite a lot of practice to be able to do this with confidence. And it is important from the psychological point of view to make it look easy—otherwise the audience are not so ready to accept this mode of vibration as something that is a natural to the string rather than as some trick of the demonstrator.

Once this has been done it is relatively easy to obtain other modes by oscillating at approximately twice, three times, etc, the fundamental frequency, and it is usually possible to obtain at least the fifth mode. But it is rather like playing a musical instrument—it is no use trying to force the rope, you have to be sensitive to its natural frequencies and adjust very rapidly until the desired one is found, when quite a large amplitude can be maintained with very little movement of the hand holding the end. Again there is no substitute for lengthy practice.

3.6.3 Evolution of style of presentation

In this category I will instance the evolution of my particular way of presenting the classic demonstration that was first done by Chladni (1756–1827) at the very beginning of the nineteenth century, and will begin with Chladni's own description of its original evolution.

Demonstration 3.6

Tyndall[62] quotes Chladni's account as follows:

> *. . . I noticed that the science of acoustics was more neglected than most other portions of physics. This excited in me the desire to make good the defect, and by new discovery to render some service to this part of science. In 1785 I had observed that a plate of glass or metal gave different sounds when it was struck at different places, but I could nowhere find any information regarding the corresponding modes of vibration. At this time there appeared in the journals some notices of an instrument made in Italy by the Abbé Mazzocchi, consisting of bells to which one or two violin bows were applied. This suggested to me the idea of employing a violin bow to examine the vibrations of different sonorous bodies. When I applied the bow to a round plate of glass fixed at its middle it gave different sounds, which,*

> *compared with each other, were (as regards the number of their vibrations) equal to the squares of 2, 3, 4, 5, &c.; but the nature of the motions to which these sounds corresponded, and the means of producing each of them at will, were as yet unknown to me. The experiments on the electric figures formed on a plate of resin, discovered and published by Lichtenberg, in the memoirs of the Royal Society of Göttingen, made me presume that the different vibratory motions of the sonorous plate might also present different appearances, if a little sand or some other similar substance were spread on the surface. On employing this means, the first figure that presented itself to my eyes upon the circular plate already mentioned resembles a star with ten or twelve rays, and the very acute sound, in the series alluded to, was that which agreed with the square of the number of diametrical lines.*

He may not have succeeded in his aim to 'make good the defect' because in physics education sound still tends to be a Cinderella subject, but he certainly made a niche for himself in the annals of lecture demonstration, though, as in many other topics, it was the rather shy Wheatstone who provided the basis of a theory.

There are, of course many different ways of performing the Chladni experiment; the vibrations can be excited in a metal plate by means of an electromagnetic driver fed with a variable-frequency sinusoidal current, or a wooden plate (e.g. the back plate of a double bass) can be excited by suspending it on foam pads over a large loudspeaker fed with variable-frequency sinusoidal current. But my favourite uses Chladni's original idea of bowing, but with a metal plate. My own is about 20 cm square and is fixed to a 15 mm diameter pillar by a single screw through its mid point. It is made of 1 mm thick brass and it is important that it should be cast plate, and not rolled, if symmetrical patterns are to be obtained.

When I first started to perform this demonstration I was never quite certain what pattern or note would be obtained and I would just touch the edge at various places while bowing and hope for the best. But gradually I found the trick of forcing a particular mode. For every different mode there are specific points along each edge where a nodal line intersects the edge and this is where a finger must be placed, and the bowing must occur mid-way between any two adjacent points. But unless you have several assistants to provide additional fingers it is impossible to cover every nodal intersection and there are often several patterns that have some intersections in common, and the selection of touching and bowing points needs to be made so that it is unique to one mode. For example the three modes shown in figure 3.6 all have an intersection at each of the four corners. To induce (a) on its own it is necessary to touch one corner (symmetry takes care of the other three), but then if you bow in the middle of a side there is ambiguity between (a) and (b); to remove this the bowing is done at the point marked p, thereby eliminating mode (b) which would require a node here. To obtain (b) on its own it is necessary to touch at x and y to induce the two families of nodal lines and to bow at q in order to inhibit (c) which would have a node there. Finally to obtain (c) on its own three places must be touched, x, y and z, and the bowing is at r. Figure 3.7 shows photographs of five different modes being produced.

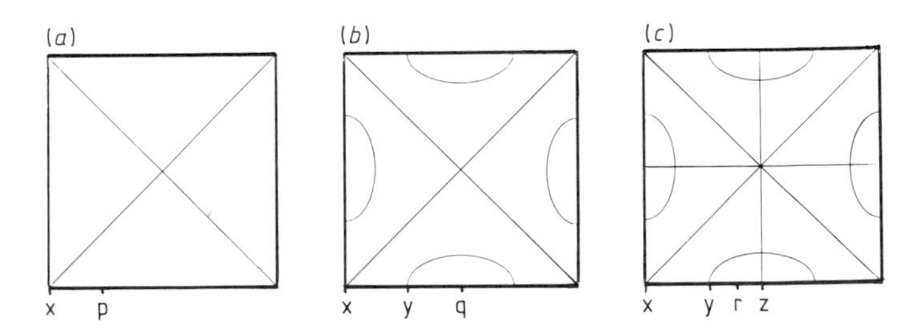

Figure 3.6 Diagram showing the way in which single modes on the Chladni plate can be induced.

Having reached the point of being able to produce several different modes at will I find that it is much more effective to predict the pattern that I am going to produce. The aim, as with the rubber rope in Demonstration 3.5, is to make it look as simple as possible so that the audience appreciates that it is a fundamental property of the plate that is being demonstrated rather than some feat of sleight of hand on the part of the demonstrator. The final step in the evolution of this demonstration (so far!) has been to find two tuning forks, one of which has a frequency corresponding to one of the well defined modes of the plate and the other which does not correspond to any mode. The note of the first fork is strongly amplified when it is struck and placed on the plate and the second is not. I now use this as an introduction to the reasons why the resonances of the front and back plates of stringed instruments are so important.

3.6.4 *Towards greater portability*

My third example concerns the evolution of a series of demonstrations on colour and colour mixing, which started off needing eight different slide projectors for the demonstrations, not counting the main projector used for illustrations, and some very large, heavy auxilliary equipment, but ended up in a portable form using only three projectors. This demonstration has now been used as far afield as the USA and Singapore.

Demonstration 3.7

One projector was used to project a large spectrum on to the screen and then to recombine the colours; two were used for an experiment to show how the colour brown arises; three projectors were used for standard colour mixing experiments with red, blue and green filters; and a further two were used to demonstrate the Land effect. (The final set of two were used because the adjustment to obtain exact overlap of the images of two slides was very delicate and was difficult to re-do in the middle of the lecture, so it was important that these two were not touched once set.)

We will begin with the spectrum experiments. Newton[63]*, describes various ways of achieving the recombination of the spectral colours on a relatively small scale and I sought a method of performing the experiment on a sufficiently large*

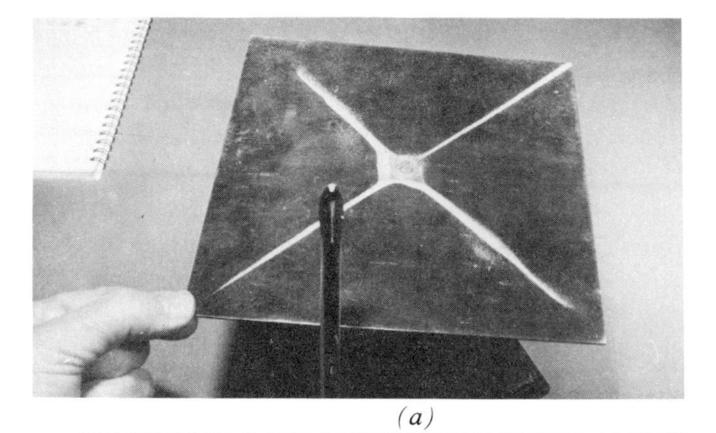

(a)

(b)

(c)

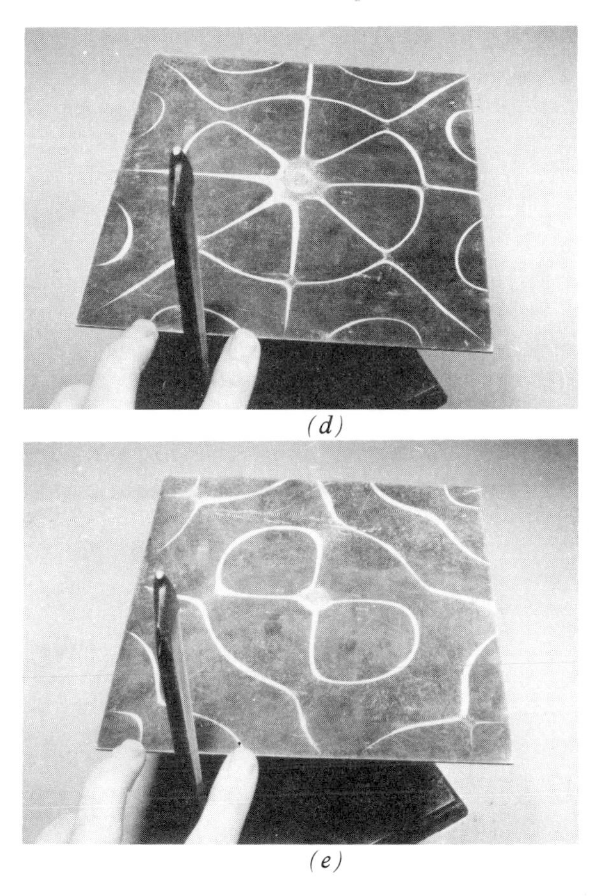

(*d*)

(*e*)

Figure 3.7 Five different modes on the Chladni plate; the respective frequencies are (*a*) 86 Hz, (*b*) 1200 Hz, (*c*) 2182 Hz, (*d*) 3828 Hz and (*e*) 2275 Hz.

scale to enable an audience to see clearly what is happening. My arrangement for projecting a large spectrum on the screen is similar to that shown in figure 2.7 (p. 70) and it is possible to achieve a very bright, relatively pure, spectrum about 1 m in length and 0.4 m high using a 300 W tungsten–halogen slide projector and a 1 mm slit. The slit width has to be a compromise, a narrower slit gives a purer spectrum but less light.

It seemed to me that a good way to recombine this so that the audience could have no doubt what was being done was to allow this spectrum to fall on a large (1 m diameter) searchlight mirror which would then produce a small patch of light of mixed colours. But the mixing was incomplete and the patch too small to be seen by the audience. So the next stage was to allow this patch to fall on an opalescent sphere of glass about 0.2 m in diameter of the type used as a globe surrounding a light bulb. The patch falls on the rear of the sphere and is then scattered and the whole sphere is illuminated with the mixed colours (figure 3.8). If the whole spectrum is used a very good approximation to white is achieved and, in a darkened room, it is quite striking. If various elements of the spectrum are successively eliminated the corresponding negative colours can be demonstrated and the cyan (minus red) and magenta (minus green) achieved in this way give very beautiful effects on the sphere.

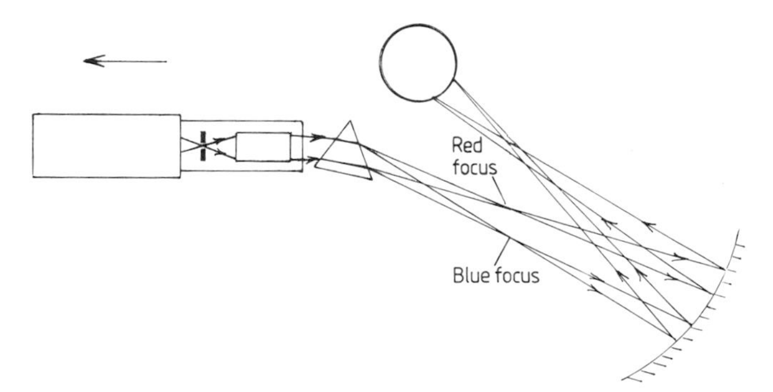

Figure 3.8 Using a large searchlight mirror to recombine the spectrum.

However, the mirror is large and heavy and therefore not easy to transport. Hunting for an alternative I suddenly realised that the large mirror is totally unnecessary. All that is needed is a small plane mirror placed quite close to the prism (figure 3.9) and to readjust the focus of the projector so that the spectrum is in focus on the sphere; the mixing to give white will occur even if it is out of focus, but elimination of certain colours from the spectrum can only be done if it is sharply focused.

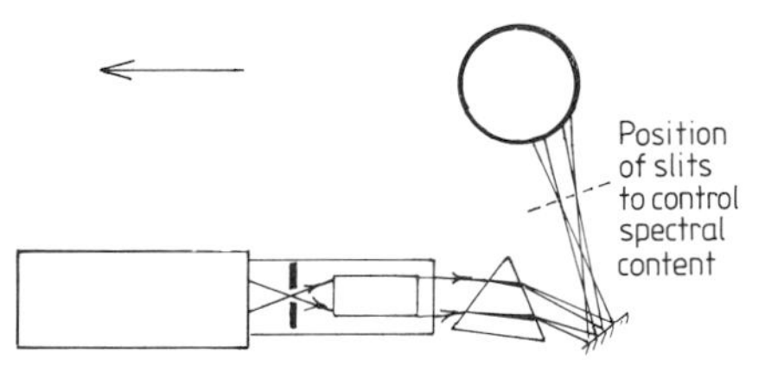

Figure 3.9 Using a small plane mirror to recombine the spectrum.

I then realised that one of the three projectors used for colour mixing could be used for these experiments and so the number of projectors was reduced to seven.

For most of the experiments using projectors some sort of dimmer is needed to control the brightness of the illumination on the screen; I have found that the small thyristor dimmers now sold for domestic use in room lighting are very effective (the projector fan is slowed down but in practice the speed seems to be adequate at each reduced output level).

The 'brown' experiment consists of using two projectors, one with a slide of a black square on a white ground and the other with a white square on a black ground. The images are carefully adjusted so that they coincide exactly and then, using dimmers on the projectors, the illumination of the inner square can be controlled separately from that of its surroundings (see figure 3.10). A patch of yellow paper exactly the same size as the image of the square is then placed on the screen and can be illuminated at any desired level compared with its surroundings. It is quite difficult to obtain absolutely precise coincidence of the two squares and a black border round the yellow square improves the appearance enormously. Keeping the surroundings bright and dimming the square produces the appearance of brown and with care it can be matched to a genuine brown square placed in the surrounding area. But then dimming the surroundings returns the yellow appearance to the patch. Dimming both patch and surroundings together provides a very good illustration of colour constancy under different levels of lighting. The problems of adjusting three projectors for colour patch experiments, or the two projectors for the brown experiment and the Land experiment are that push-through slide holders prevent them being placed very close together and do not locate the slide accurately enough in the gate whereas magazine-type projectors are not easy to operate with only one or two slides, and again are usually so wide that the projection lenses cannot be placed very close together.

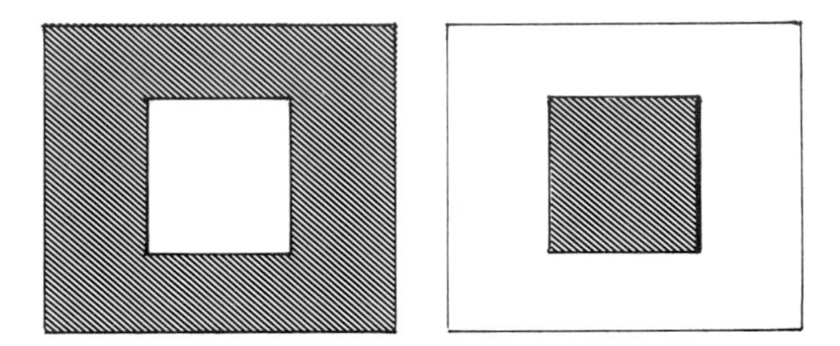

Figure 3.10 The two slides used for the 'brown' experiment.

The solution to this problem, which also enabled me to reduce to only three projectors, was to use 300 W Halight projectors, which normally have push-through slide holders but which can easily be modified to take only one slide, inserted from the top, and therefore locating very precisely in the gate (figure 3.11). Three of these projectors mounted close together, side by side, on a home-made mount that enables each to be tilted in all three directions so that exact alignment of three slides can be achieved, could then be used for all the experiments. The centre projector is used for the spectrum and recombination, all three are used for colour patch experiments, the centre and right-hand one are used for the Land experiment and the centre and left for the brown experiment. (This allows independent fine adjustment of the left- and right-hand projectors; the colour patch experiments do not require such great precision.) Since then other experiments have been added using the same three projectors, and the evolution that was forced by the needs of transport has, in fact, resulted in great improvements in performance, facility of use and in the variety of possible demonstrations.

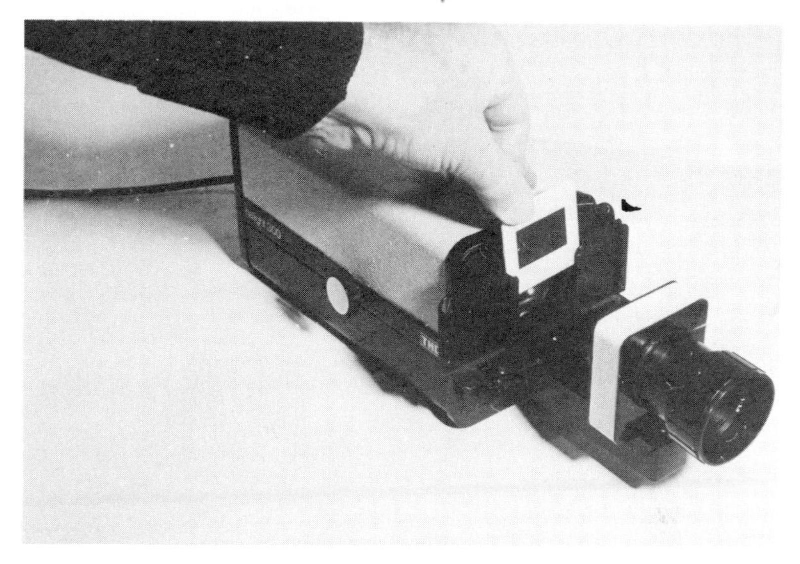

Figure 3.11 The modified Halight projector with fixed slide holder loaded from the top.

In §3.2 I mentioned the use of a rack to hold the slides used for this demonstration; they were shown in figure 3.2 and are numbered in the diagram of figure 3.12. The complete sequence of demonstrations possible with this one set of three projectors and thirteen slides is as follows.

(1) Slide 1, slit for spectrum and recombination.

(2) Slides 2, 3 and 4 blue, red and green discs for Young's colour patch experiment.

(3) Slides 5, 6 and 7, full-size blue, red and green filters for colour mixing, for illuminating coloured fabrics with different mixtures of coloured light and for performing the 'magic transformation' experiment in which a drawing in red ink is superimposed on one in green ink (e.g. Dr Jekyll in red and Mr Hyde in green) and the transformation from one to the other is effected by cross fading from green light to red; if the three projectors are adjusted in brightness to give a white patch on the screen, coloured shadows in yellow, magenta and cyan will be thrown on the screen by any object, e.g. a hand, placed in the beams.

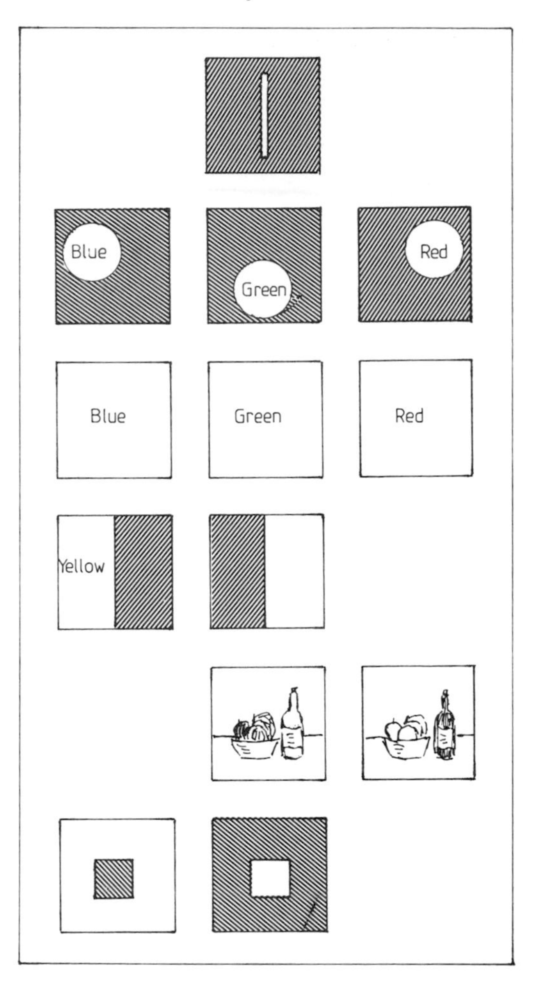

Figure 3.12 The arrangement of the slides for colour demonstrations on the rack of figure 3.2(*a*).

(4) Slides 8 and 9. Slide 8 has the left-hand half yellow and the right-hand half black and slide 9 has its left-hand half black and its right-hand half clear. The projector carrying slide 9 has its lens covered with a mercury yellow interference filter. The result on the screen is two patches side by side which both appear yellow and can be matched in brightness using the

dimmers. Objects placed in the beam from the yellow slide look relatively normal, whereas those in the beam from the interference filter appear in monochrome (as do objects under sodium street lighting). The most dramatic way of concluding this section of the demonstration is to place my own face first in the beam from the yellow slide and then in the monochromatic beam. Audiences of all ages enjoy seeing the lecturer looking ill!

(5) Slides 10 and 11 are used for the Land effect. Slide 10 is a picture of a still-life scene taken through a green filter and slide 11 is one taken through a red filter. If placed in two projectors, with red and green filters being placed over the corresponding lenses, a reasonable colour reproduction of the original scene is produced even though no blue is included. If now the red filter is removed so that one projector has no filter and the other a green filter and the brightness is adjusted with the dimmers to compensate for the absorption of the filter a surprisingly good colour reproduction including reds and yellows results. This is the essence of the Land effect[64]*.*

(6) Slides 12 and 13 are the two slides used for the brown experiment described above.

3.6.5 Towards cheaper equipment

Two further stages of evolution have occurred with some of these demonstrations in an attempt to find first a way of making them even more transportable and finally a way in which schools that cannot afford three projectors and the dimming equipment could still perform some of the experiments.

Demonstration 3.8

This began when I was looking for something a little less fragile than the glass globe for the spectral recombination experiment. It occurred to me that a cube would be as effective as a sphere for mixing and that a translucent window would be all that was needed so that the rest could be opaque. The result was a box made of stiff card with sticky-tape hinges that could be folded

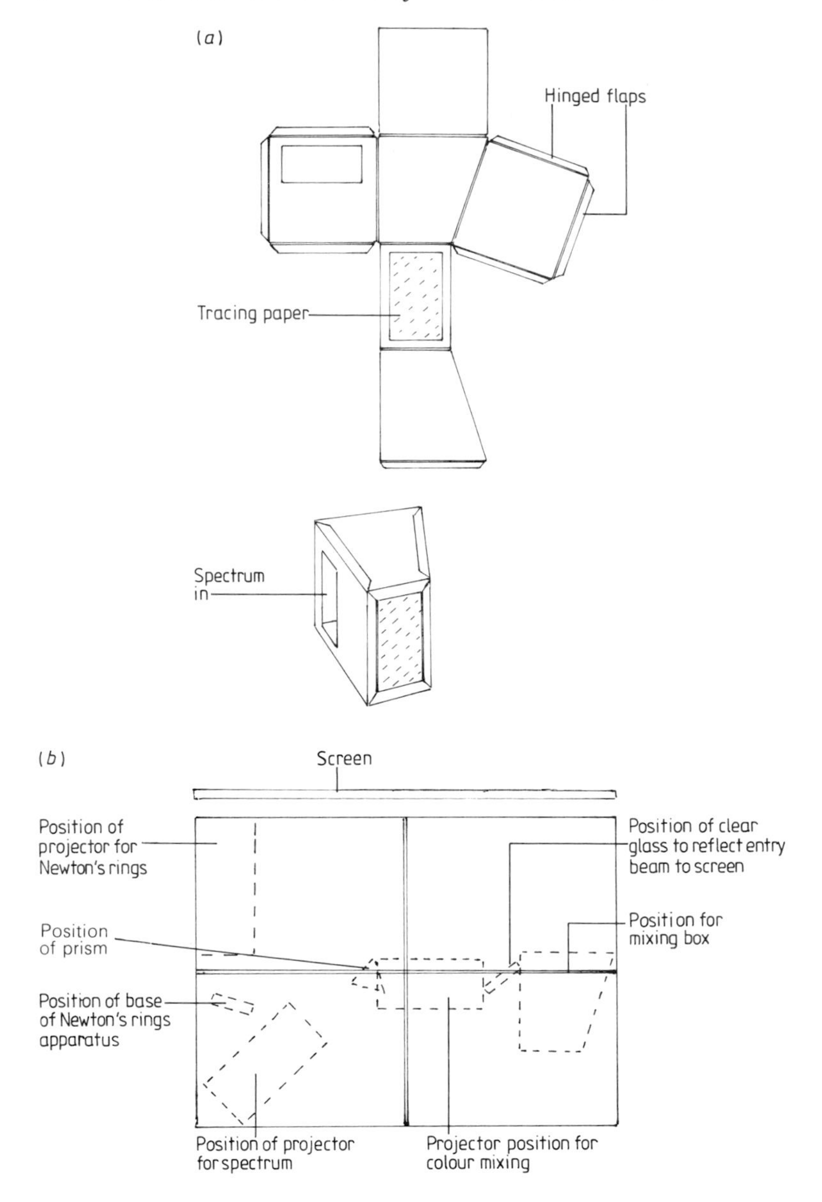

Figure 3.13 (*a*) Construction of the portable colour mixing box; the departure from the perfect cube increases the efficiency of the mixing. (*b*) Portable card which folds into four and assists in rapid location when the single projector is used for several different demonstrations.

flat for travelling; figure 3.13(a) shows the design. The spectrum enters through the hole, is mixed by the box and illuminates the translucent panel with the mixed light. If only one projector was to be used it would obviously have to be placed in different orientations and so I produced a large card (800 × 1120 mm) which would fold into four for packing and on which the outline of the position of various items could be drawn. Figure 3.13(b) shows a typical layout. Portions of the spectrum can be cut out by means of strips of cardboard supported in the beam just before it enters the mixing box.

I then realised that the same box could be used for colour mixing experiments. A special slide consisting of equal strips of red, blue and green filter was made. When inserted in the top-loading slide slot of the single projector strips of opaque material could be slid up and down over the filters to vary the proportions of the three colours (thus eliminating the need for dimmers) and the resultant triple beam is shone into the mixing box as for the recombined spectrum. Colour mixing and such concepts as negative colours can be demonstrated quite easily with this arrangement.

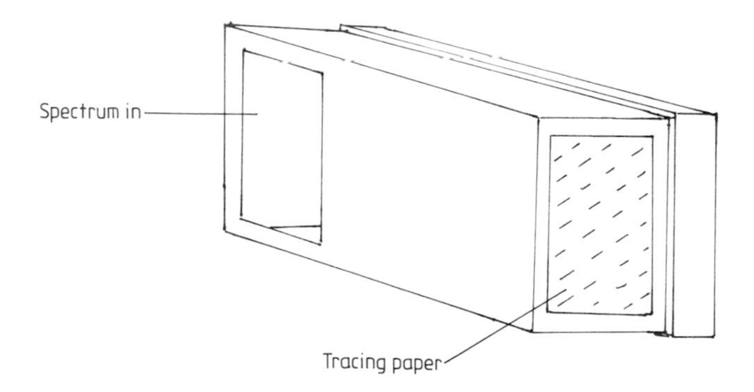

Figure 3.14 Simple version of the colour mixing box using a shoe box.

The final evolutionary stage was to simplify the colour mixing box down to the shoe box shown in figure 3.14 which works well for teaching small groups. The box, which is sprayed matte white inside, has two holes cut in it one of which is covered with tracing paper. Either the spectrum or the three-coloured beam can be shone into the hole and the resultant mixed colour observed as a luminous glow on the tracing paper.

Another demonstration used in my colour lectures has already been described in its original form as Demonstration 3.1 In that form it was extremely bulky, but its evolution followed an interesting path not just because of the need to cut down the size for travelling.

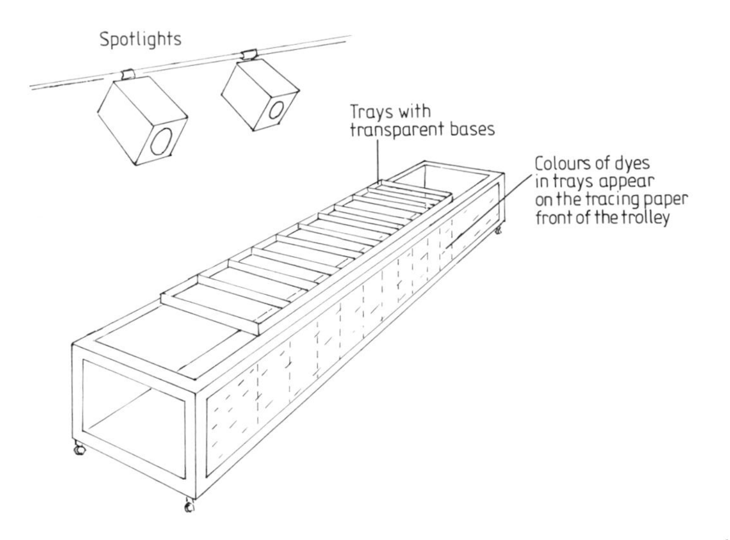

Figure 3.15 Trolley with trays and overhead spotlights for use when the audience is largely below stage level, as, for example, in a school hall.

Demonstration 3.9

I discovered that one particular theatre at which I was to present the dye-mixing demonstration (Demonstration 3.1) had a stage that was so much higher than the audience that only those in the balcony would be able to see the dyes in the trays. Nevertheless I still wished to retain the dramatic impact of the large quantities of dye (figure 3.3, p. 110). So we built a trolley (figure 3.15) and used trays with transparent bottoms. A piece of translucent tracing cloth was draped across the front and spotlights were arranged to cast obliquely through the trays on to the cloth so that, as the dyes were poured in, the dye colour could be seen quite brilliantly. The concentration of the dyes had to be increased, but the result was even more spectacular than the original. Then it became clear that the apparatus for this lecture would have to be slimmed down to make it more portable so some smaller, clear Perspex trays, holding about half a litre of dye each, were obtained and the buckets replaced by plastic bottles. In most lecture theatres the rake is sufficient to allow the audience to look down on the top of these dishes, but they were about ten centimetres deep and so, with a white card behind, could be viewed from the front in a flat theatre (figure 3.16).

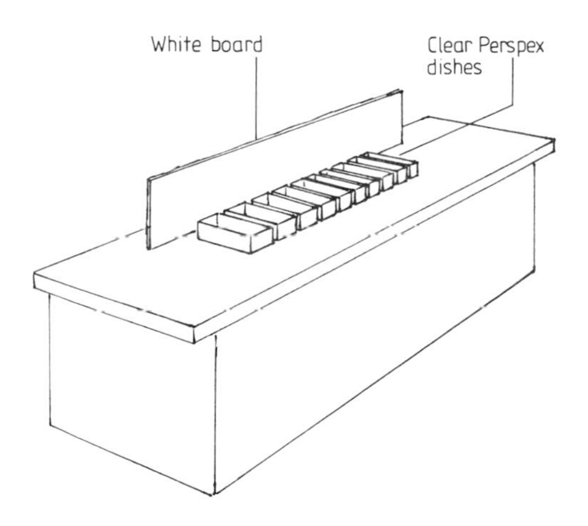

Figure 3.16 Small, portable, transparent trays for use on the bench when the audience is level with the bench or above it.

However, when I arrived at one particular lecture theatre the lecture bench was so high that most of the audience could not see the colour of the dyes even through the fronts of the trays. The resources of the university were mustered and three overhead projectors were produced so that the colours of the dyes could be seen on the screen. The effect was so interesting that it set me thinking about an even more compact and univeral way of presenting this demonstration. I now use a set of 50 mm diameter clear plastic petri dishes that are mounted in a wooden board (figure 3.17) (see Demonstration 2.9, p. 83). The concentrations of the dyes had to be changed yet again, but the result is very effective, the weight and size of the containers has been drastically reduced, the quantity of dyes needed to be transported is relatively minute and only one projector is needed.

Figure 3.17 Wooden board with petri dishes for colour mixing on the overhead projector.

3.7 COPING WITH DISASTERS

3.7.1 Introduction

One of my recurring nightmares, though I am glad to say it has not come true yet, is that I shall arrive for a lecture with the wrong set of apparatus. The nearest I came to this was only a year or two ago when I was booked to give a lecture on 'Colour' about three weeks after a lecture visit to the USA. Through some mischance over documentation my apparatus had returned by air freight to London Airport but could not be released from the customs authorities until the mistake had been rectified. Fortunately I was able to give my hosts two or three days' notice that the lecture would have to be on a different topic!

But I suppose if I did ever turn up with the wrong apparatus I would have to offer the audience the alternatives of a boring lecture on the advertised topic with no demonstrations or a demonstration lecture on the wrong topic.

Disasters of one sort or another are bound to happen and I suppose the most generally useful piece of advice I can give is don't lose your nerve; the audience do not know what you intended to do or what should have happened, so the chances are that only you will realise the extent of the problem. One of the most shattering experiences I have had was the very first time I gave a discourse at the Royal Institution; it was on the theme of 'Physics and music' and I intended to finish with a piece of music played on the tape recorder and with its waveform displayed on a cathode-ray oscilloscope. It is a tradition at the Royal Institution to try to finish the discourse precisely as the clock strikes ten and I had thought of using the favourite broadcaster's trick of starting a two-minute tape at two minutes to ten with the volume turned to zero. I could then make my concluding remarks, turn up the volume and finish exactly at ten. But, for the only time in forty years or so of demonstration, when I pressed the button to start the recorder, the tape broke; I had to rethink the ending with two minutes to go. But I discovered later that, though I was in a complete state of confusion, the only people who realised that anything was wrong were my wife and the man who was assisting me, Mr Coates!

3.7.2 Disasters from outside causes

Disasters sometimes arise for reasons beyond the control of the lecturer or of the hosts. But in my experience (perhaps I have just been lucky) it is surprising how often some ingenious improvisation can resolve the problem. On one occasion I had arrived at a residential boys' school in the heart of Nigeria, had set up my apparatus, watched the audience of 500 file in, and was about to be introduced by the chairman when the power failed, immobilising my equipment and plunging the whole room into semi-darkness. I was rapidly trying to work out how I could give the lecture with only a few purely mechanical demonstrations when the head boy came up and told us that the headmaster had a petrol-driven stand-by generator but that he was away from the school on leave. A party of boys sprinted across to his house and returned shortly with the generator and I was able to start the lecture only 15 minutes late. Since that experience I have spent a little while planning a lecture with only mechanical demonstrations just in case that should happen again!

3.7.3 To apologise or not?

Of course, if an accident of some kind occurs that has a direct impact on the audience (e.g. the escape of some malodorous gas or a cloud of smoke) there must be a reference to it and an apology of some kind. But, as is so often the case, Michael Faraday[65] has something to say on this topic:

> In lectures, and more particularly experimental ones, it will at times happen that accidents or other incommoding circumstances will take place. On these occasions an apology is sometimes necessary, but not always. I would wish apologies to be made as seldom as possible, and generally only when the inconvenience extends to the company.
>
> I have several times seen the attention of by far the greater part of an audience called to an error by the apology which followed it.

Occasionally it can happen that you lose the thread of an argument, or actually make an incorrent statement. In that case a simple apology and a restart is all that is necessary.

3.7.4 Turning disaster to advantage

Sometimes a mistake, or a demonstration that does not work, can actually enhance a lecture—but only if it occurs spontaneously; it should never be deliberate. I suppose it works on the same principle as that of the trapeze artist who falls off the first time he attempts a trick and gets a much enhanced round of applause when he succeeds the second time. One classical example of this occurred during the recording of one of the 1971–2 Christmas Lectures. In one lecture, which was all about the interactions of technology and music, I used a bank of synthesisers, which were then brand new. My intention was to synthesise a few bars of a Bach Two-part Invention live, in order to demonstrate the principle, and in particular the sequencer, or memory, which, because it stored control signals rather than waveforms, could play back at any speed without change of pitch and with any desired timbre. The plan was to play the first part slowly with one finger on the keyboard, then to use the sequencer system to play this back and add the second part with a different timbre, and then to play both parts back together and to speed it up to normal tempo. I succeeded in storing both parts and then instead of pressing the replay button I pressed the erase button, instantly destroying the whole piece. I quickly explained to the audience and then repeated the whole thing, this time successfully. I assumed that the television producer would edit out the mistake—but he insisted that it made much better television to leave it in, thereby enhancing the dramatic impact of the final result.

3.8 DEALING WITH DIFFERENT AGE GROUPS

3.8.1 Different approaches

One of the outstanding advantages of a demonstration lecture over straight talking, with or without visual aids, is that it can ease enormously the problem of an audience of very mixed age groups or abilities. But before discussing that aspect further it will be helpful to look at the differences in approach that are needed even when dealing with relatively homogeneous groups.

In my experience one of the most difficult audiences to face is the one composed of experts in the field. This may be a surprise to you; the reason is that even experts have different rates of assimilation and if you pitch the standard for the middle rate there will be an appreciable fraction of the audience who are left behind. If you pitch the standard lower then some of the audience are bound to feel that you are underestimating them. However, if there are just a few people present who are not experts, the lecture can be pitched a little lower for their benefit and you can refer to this fact in your preliminary remarks. Honour is then satisfied; the experts at all levels assume that you are explaining things especially carefully because of the presence of the non-experts, but at the same time most of them find the slower pace very helpful!

One particular example of this problem is when I am asked to address an audience of sixth-formers on the physics of music; if I know that they are all scientists I have to be extremely careful to ensure that every item of background that I am assuming has been covered in their school courses and yet that I do not go over familiar ground. If on the other hand there are a few musicians present they can be made the excuse for starting a bit further back without insulting the scientists!

Intelligent laypeople, who form the kind of audience likely to attend a lecture billed as for the general public, are often the easiest to address. You can happily assume that they know very little about the scientific aspects of your topic but are nevertheless very ready to absorb your material. Demonstrations are extremely useful in this case because, if carefully chosen, they can provide explanations of quite difficult concepts that would be difficult to put over with words or pictures alone. An example of this is a favourite demonstration of mine about images. The basic physics that is to be explained is really that of the Abbe theory of microscopic vision or, in more mathematical terms, the notion of Fourier transformation, and it contains some quite complex ideas. But the essence of the idea can be put over to non-scientists using the following demonstration sequence; in fact, of course, the same demonstration can shed new understanding on the subject for those who are already familiar with the theory.

Demonstration 3.10

I begin by drawing attention to the fact that seeing is a process of information exchange; the audience are absorbing information about me—the pattern on my tie, the fact that I wave my hands about as I talk, etc—and I am simultaneously absorbing things about them—nobody is reading the newspaper, no one has gone to sleep, a student in the front row has flaming red hair, etc. Obviously the medium of the exchange is light; light falls on the objects in the room and is modified by the object and then collected by the eye of the observer. Clearly the information must be encoded in some way in the light beam. I then reinforce and amplify the point by the demonstration[66]. I remove the lens from a slide projector, insert a slide and, of course, the result on the screen is a totally indecipherable patch of light. But I then point out to the audience that all the information about the slide must be on the screen; obviously as soon as I put the lens in and adjust it properly we obtain an image of the slide, but equally obviously the lens cannot 'know' what is on the slide, it can only rearrange the information that is already there to make it intelligible. We could say the light falling on the slide picks up the information in coded form and the lens then decodes it.

I usually go on to explain that this is the basis of most image-forming systems; radiation of some kind (as alternatives to light it could be the radio waves in radar, the ultrasonic waves in medical scanning, the x-rays in x-ray crystallography, the electrons in electron microscopy, and so on) falls on the object and is scattered by it and the scattered waves are then recombined in some way to form the image. In the case of the slide projector, the decoding, or recombination, by the lens involves the process we call focusing. But, if you think about it, focusing merely involves making the image look like you think the object looks; if the slide is replaced by a piece of fluted glass it becomes possible to change the appearance of the image through a sequence of patterns as the focus is adjusted and there is no way of telling which one is 'right'.

This problem is at the heart of some of the fierce controversies that occurred between microscopists in the nineteenth century. Arguments occurred about the shapes, numbers of holes, etc, in the skeletons of minute sea-creatures called diatoms. In most cases the answer is that both opponents are right in their drawings of the shape of a particular specimen, but each was adjusting the focus differently.

Starting from this demonstration we can proceed to ask what is the form in which the information is encoded and go on to discuss the idea of phase relationships, holograms, coherence, etc, making use of Demonstration 2.1 (the striped string) in the process and arriving at ideas on resolution that would be very difficult to convey to non-scientists in any other way.

3.8.2 Lectures to undergraduates

I have never understood why it is so often assumed by university lecturers that their students are so highly motivated that the usual rules for retaining the attention of an audience, for making the subject exciting, for introducing elements of humour and drama can all be suspended. A great many lectures still consist of unrelieved chunks of theoretical work without even the odd audio-visual aid. There are notable exceptions, and some institutions now provide courses of preparation for new lecturers. Of course I see the need for the presentation of solid material, especially at the higher levels, but I am quite sure that even the most dedicated students can be helped if the lecturer takes note of some of the matters discussed in earlier sections such as §2.2 and 2.3. And I remain convinced that lecture demonstration can play a vital part, even if it does no more than provide a variation in tension.

It can be argued that the opportunity for the students to do experiments for themselves during laboratory periods decreases the need for demonstration. But there are two counter arguments: first, it is impossible to arrange that all the students in a class do the experiments corresponding to a particular lecture topic simultaneously when that topic is being covered, and it is undoubtedly a great advantage to have the theoretical and experimental aspects presented at about the same time; and secondly the modern tendency (with which I heartily agree) to give students rather lengthy, project-type experiments in the laboratory rather than the multitude of short experiments that used to be popular, means that the number of topics that can be covered experimentally by any one student is relatively small.

Obviously there are many topics in physics for which demonstrations are virtually impossible. But it is rare to find a topic in which some variation in tension cannot be achieved by the use of a few slides, a piece of film or videotape, or even the narration of an historical anecdote about the researcher involved. But I would still plead for the use of demonstrations wherever possible.

In my introductory lectures to first-year students I always tried to include a few demonstrations that would give a flavour of what was to come. In particular I tried to include one or two rather surprising demonstrations in order to warn against the dangers of Aristotelian thinking. It is quite surprising how many students still rely on thinking what the result of an experiment will be without actually doing it. Typical of my experiments in this category would be the use of a solution of aluminium stearate in a high alcohol which appears to have the consistency and colour of golden syrup, but which is a non-Newtonian liquid. If poured from one beaker to another it seems to be behaving like golden syrup until you cut the stream with a pair of scissors, when the upper part jumps back into the beaker. But what of more serious demonstrations? In §2.1 I defined three types of demonstration and I propose now to give two examples of undergraduate demonstrations in each.

Demonstration 3.11

The first example of an unconventional visual aid is the use of a clock spring to illustrate the behaviour of the phasor diagram for diffraction at a single slit. The use of phasor diagrams is discussed in many optics textbooks (e.g. Taylor[67]) but I have often found that students have difficulty in visualising the behaviour of the components as the viewing point moves off the axis. A piece of old clock spring can be used effectively to represent the sum total of the components when stretched out flat (representing the amplitude at the central maximum) and then allowed to curl up as the viewing point moves off axis. It is particularly useful in showing how the minima in the pattern are regularly spaced, but the maxima are displaced slightly from the midway positions between the minima.

Demonstration 3.12

The second example of an unconventional visual aid is the use of a model to demonstrate interference when waves are out of phase and also the concepts of wave and group velocity. The model (figure 3.18) consists of a set of vertical rods on each of which slide two wooden beads; the beads are separated by a length of drinking straw and the lengths are cut so that the upper beads lie on a sine curve. Two templates can be slid in to raise the lower beads. One template has the same wavelength and amplitude as the wave represented by the upper beads and as that is slid in the relative phase of the two components changes and so the result is that the top beads vary from a sine wave of twice the amplitude to a straight line. The second template has a slightly different wavelength and when that is slid in the upper beads show the beat phenomenon. The beat maximum can be regarded as a group of waves and if the model itself and the template are moved along the bench at different speeds the phenomena of wave and group velocity can be demonstrated.

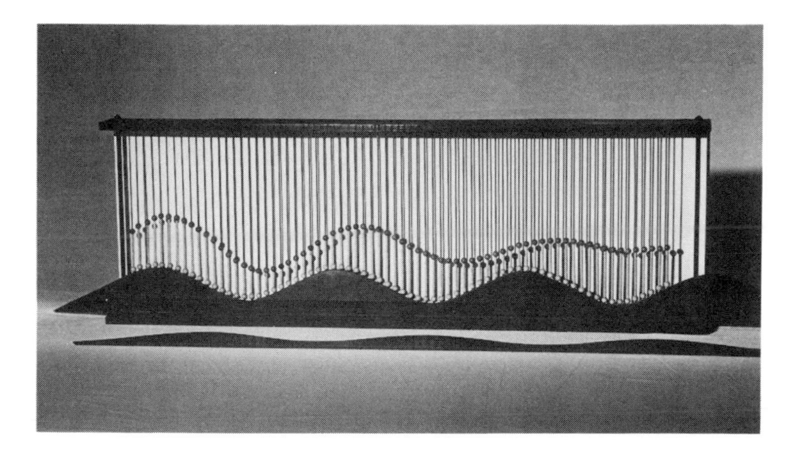

Figure 3.18 Model for demonstrating addition of two sine waves. The upper (dark) beads are separated from the lower (pale) beads by lengths of drinking straw sliding on wires. The straws are cut so that when the template is not in position the dark beads display a sine wave. The template that has been inserted has a sine wave profile but of slightly shorter wavelength than for the beads and hence a 'beat' waveform is produced with the maximum at the left and the minimum at the right. If the template lying on the bench, which is of the same wavelength as for the beads on their own, is inserted, the effect of their relative phase on the resultant can be demonstrated to give any amplitude between zero and twice the straw amplitude thus illustrating interference.

In the category of analogue demonstrations I have chosen the elegant microwave analogue demonstration of the rotation of the plane of polarisation by dextro, laevo or racemic liquids described by Foxcroft[68] and the well known Vinycomb wave model[69].

Demonstration 3.13

Microwave equipment for demonstrating all kinds of wave phenomena, such as interference, Döppler shift, etc, has been available from suppliers of science-teaching apparatus for many years, but I think one of the most exciting demonstrations is that devised by Geoffrey Foxcroft on the rotation of the plane of polarisation. Sheets of Perspex have embedded in them small helices of copper wire all wound in the same sense and arranged parallel to each other in a regular array to represent molecules on a crystal lattice. When microwaves pass through this their plane of polarisation is rotated, as can easily be shown by the need to rotate the detector. But he points out that certain liquids rotate the plane of polarisation as well as crystals and this is perhaps more difficult to understand. However, the analogue still works. Copper helices are embedded in polystyrene balls and thrown together in a box, thereby representing the orientations that would occur in a liquid, and sure enough, if all the helices have the same sense, rotation of the plane of polarisation occurs; if half the balls are of one sense and the other half of the other, there is no rotation—the racemic form!

Demonstration 3.14

The wave model devised by T B Vinycomb[69] *is, in my opinion, one of the most successful analogue demonstrations ever devised. It is shown in figure 3.19 and consists of a row of wooden laths pivoted about their centres, but connected together by means of springs or elastic cords so that the motion of each is transmitted to the next. The ends of the laths display beautiful waves and the model can be used to illustrate many different phenomena such as reflection at free and fixed ends, standing waves, the principle of superposition by transmitting a disturbance in one direction through a disturbance travelling in the other, and many others.*

In the category of real phenomena I have chosen Fresnel diffraction, and the well known Barton's pendulums.

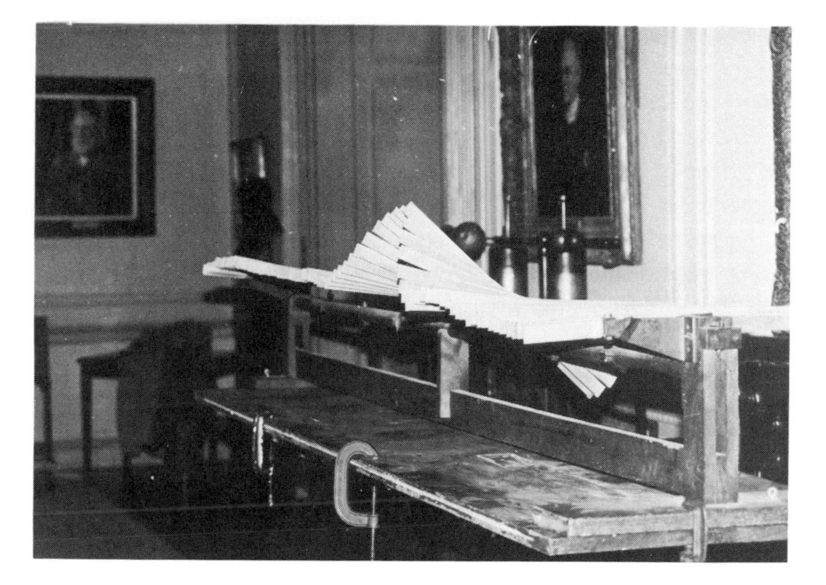

Figure 3.19 The Vinycomb wave model (Royal Institution).

Demonstration 3.15

Fresnel diffraction is a very beautiful and fundamental phenomenon that is often not demonstrated to students because it does present some difficulties. However, I have found that there is a way to demonstrate to a whole class if a super-pressure, compact-source mercury vapour lamp can be obtained. The lamp takes quite a long while to reach its full operating pressure (and hence minimum size of arc) and, if switched off, must be allowed to cool before being restarted, so that it can be half an hour or so before the demonstration can be resumed. I therefore switch the lamp on well before the lecture and leave it in position at the very back of the lecture theatre with a dark screen in front of it. To perform the demonstration the screen is removed, after warning the class not to turn to look at the source or they will not be able subsequently to see the diffraction phenomena. Then I usually start by holding my arm in front of the lamp so that its shadow appears on the projection screen in front of the class. Moving the arm closer to the lamp increases the magnification, especially of the hairs, and as they grow in size, Fresnel fringes appear round them and the bright line down the centre. Keys, hairgrips, paper clips, etc, provide excellent objects and when the lamp is thoroughly warmed up the patterns can be seen simultaneously by quite a large class. However, it is also useful to let the class walk up to the screen to examine the fringes more closely. An alternative is to allow the fringes to be formed on a ground-glass screen and to display them by closed-circuit television.

Demonstration 3.16

The phenomena of forced vibration and resonance play an important part in many aspects of physics and the demonstration due to E H Barton[70] *provides an excellent back-up to the mathematical analysis that is usually performed. The arrangement is shown in figure 3.20. It consists of a horizontal cord from which are suspended a number of pendulums of different length, each having as its bob a light paper cone. A single pendulum with a heavy bob is also suspended from the horizontal cord and is arranged to be of the same length as one of the other pendulums. When the heavy pendulum is set swinging it sets up forced vibrations to the others and the relative phases can be studied. The damping of the light pendulums can be changed by adding split metal rings to the cones and hence the dependence on damping can be explored and, if the weights are considerably increased, the very sharp resonance of a single pendulum can be shown.*

3.8.3 The 7–11 age group

One of my favourite age groups is 7–11-year-old primary school children. They are bubbling with enthusiasm, anxious to help with experiments and full of totally uninhibited questions that sometimes make me think very hard. I can honestly say that it is very rare for me to lecture to this age group without learning something for myself. The questions are the equivalent of lateral thinking and often jog you out of a comfortable groove that you may have been travelling along for many years!

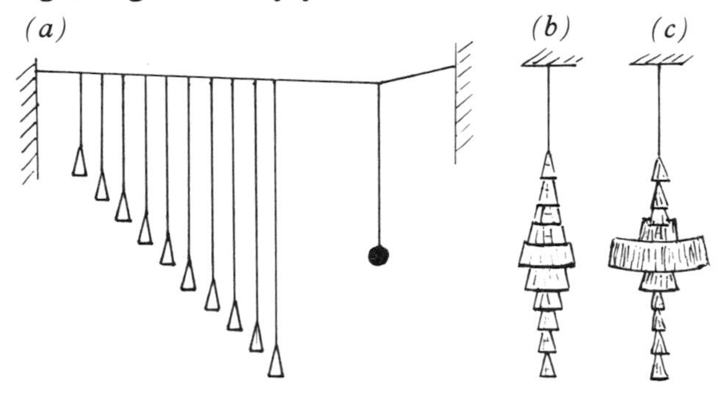

Figure 3.20 (*a*) Barton's pendulums. (*b*) End-on view showing amplitudes produced by paper cones alone. (*c*) End-on view showing amplitudes when brass rings are added to the cones; the increased mass effectively reduces the damping.

But of course 'lecturing' to this age group requires a totally different technique, and I include the inverted commas deliberately because the children play as big a part as the lecturer. Looking back to my first ventures in this field I feel that, having no training in school teaching, I had a bit of a nerve to enter the field at all. It really happened by accident. I was asked to lecture to the junior part of a comprehensive school (11–14) and then about two days before the event there was a telephone call to ask if I would be prepared to include the 6–11 age group from the primary school that happened to share the same campus. I felt that this was really too difficult a mixture and eventually agreed to take 9–14 and to give a short talk to the 6–9 group.

My opening gambit eventually proved to be my salvation, though its reception gave me quite a surprise. I started off with a rhetorical question: 'Have you ever wondered how the sound is made in a musical instrument?' And 100 children answered in chorus: 'Yes!'. Having recovered from the surprise I realised that this was the clue to dealing with audiences at this age and since then have always based my demonstration events on question and answer. A typical opening of a talk about science and music might run like this:

Demonstration 3.17

L. *How many of you are scientists?*
A. *Maybe one or two hands go up, sometimes none.*
L. *How many of you have never, ever, asked a question?*
A. *No hands go up.*
L. *Well then you must all be scientists because that is really what scientists do—ask questions, and often the answer makes them ask more questions; and of course doing an experiment is really a way of asking a question. (This can be elaborated in various ways.)*
Today we are going to ask questions about music.
Can anybody tell me what music is?
A. *Various suggestions: nice sounds, sounds you can dance to, a sort of noise, etc.*
L. *Well what is a sound?*
A. *Various answers: a noise, something that makes you jump (!), etc, but usually someone can be led to say 'something you hear'. We then go on to talk about our ears as pressure ('squash') measurers, a microphone as another pressure measurer and an oscilloscope as a way of drawing graphs. This is the opportunity for participation by talking, singing, playing a recorder, etc, and looking at the graphs of the vibrations ('wobbles').*

And so we progress by question, answer and participation in experiments.

3.8.4 Dealing with difficult concepts

There are some concepts that are quite difficult to put over to groups of young children, but which are so relevant to some common experience that a way just has to be found. A way of dealing with one aspect of conservation of momentum will be given in the section on audience participation as Demonstration 3.21. But to illustrate the point here I shall use the problem of the rainbow or spectrum which is interesting to children in its own right, but is also vital if any progress is to be made in talking about colour. Here is my solution so far (though it is still very much in the process of evolution).

(*a*)

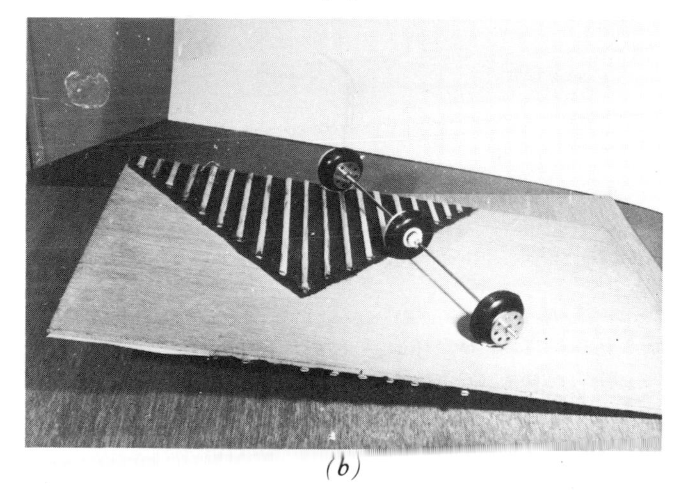

(*b*)

Figure 3.21 Strips of wood stuck on to a board to simulate a ploughed field. (*a*) When the pair of wheels is rolled down the slope it slows down visibly on reaching the 'field'. (*b*) When the pair of wheels crosses the triangular field (simulating a prism) the pair of wheels slews round, because one side is slowed down for a longer period of time than the other.

Demonstration 3.18

The first step is to introduce the idea of refraction; I use a board which has thin strips of wood glued across half of it to represent a ploughed field next to a concrete playground (figure 3.21(a)). A pair of rubber-tyred wheels on a long shaft can be rolled down the sloping 'field' and it is obvious to the audience that it slows down as it reaches the furrows. This is my model of light entering a denser medium. On the other side of the board is a set of strips arranged within the outline of a triangular prism (figure 3.2(b)) and when the roller travels down this side of the board it is deviated, as is a light beam passing through a prism. It is not too difficult then to elicit by questioning the idea that the more the roller is slowed down, the more it will be deviated.

The next stage is to introduce a glass tank on the base of which is a glass prism (figure 3.22), fill it with smoke and allow a laser beam to enter the tank, first to one side of the prism and then hitting the prism; the deviation of the brilliant red beam through perhaps 45° is most impressive.

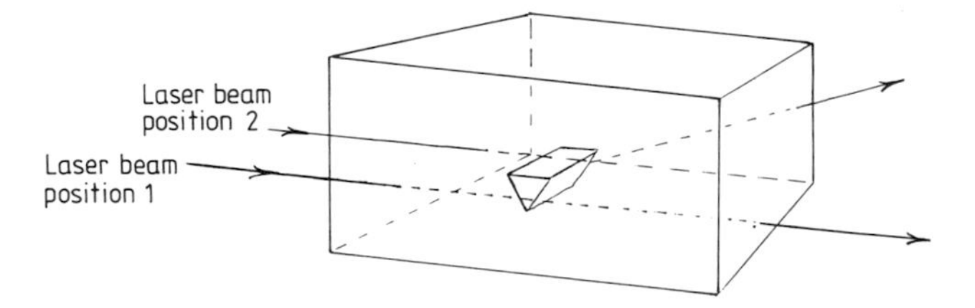

Figure 3.22 Smoke-filled tank and laser to show deviation of beam by prism.

Stage three is to introduce the fact that different coloured light beams travel at different speeds in a medium and hence, linking back to the roller, the slower ones will bend more than the faster ones. I then turn on the projector with slit and prism to produce a spectrum as described in Demonstration 2.5, but with a red filter in position so we only see the red part of the spectrum. A volunteer marks the position of the red patch on a piece of card over the screen and I then tell the audience that blue light travels more slowly than red in the glass and we take a vote on whether the blue patch will be to the left or to the right of the red. Then of course we do the experiment and another volunteer marks the blue position. We then try to guess where the green will be—all this time it is essential only to show one colour at a time. Finally we remove the filter and the spectrum is seen in all its glory—and I have never yet failed to produce a loud gasp of amazement from a young audience.

Of course prisms are not common in primary schools and so at this point I show an experiment that goes back at least to a nineteenth-century publication from which the illustration of figure 3.23 is taken. I use a simple slide projector with a 1 mm horizontal slit made with sticky black tape on a slide mount in the slide holder. An ordinary washing-up bowl filled with water and a mirror tile of the type sold for fixing to walls completes the equipment and a splendid spectrum can be projected on to the wall behind the person holding the projector. The geometry is such that it is even curved like a rainbow.

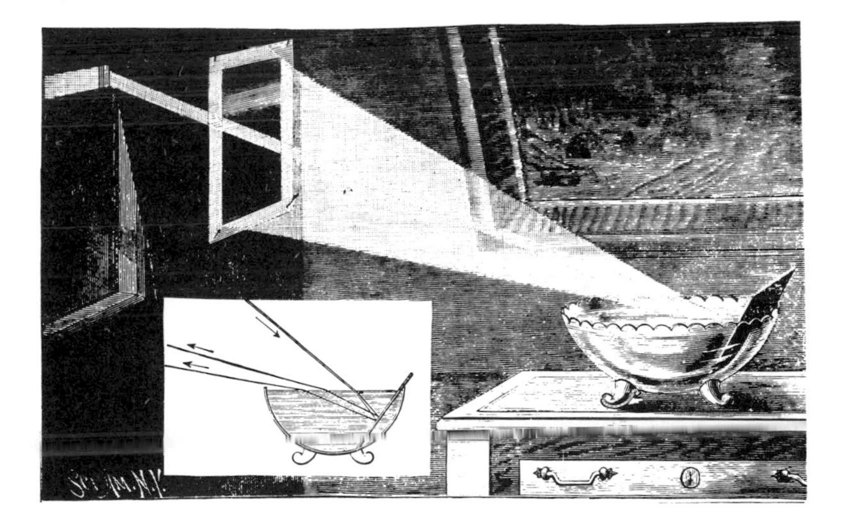

Figure 3.23 A Victorian illustration of the water prism and (inset) of the principle of its operation. (From Hopkins 1900 *Experimental Science* (London: Spon) p. 216, figure 220.)

3.8.5 Dealing with mixed audiences

But what happens if the audience is entirely mixed? I have more than once arrived to give demonstration lectures that were billed as 'public lectures' to find the age of the audience ranging from 5 to 75, though I admit that below 7 can be quite difficult. The secret seems to me to be in trying to keep the lecture going at at least three levels simultaneously.

There should be something to see or hear, possibly just a slide or transparency on an overhead projector, but, preferably, a demonstration, every two or three minutes. Then even though they may not understand a great deal, even the youngest children are entertained. The same principle holds for the non-scientists in the audience who are afraid of science and possibly have a mental block about it resulting from experiences at school; they can be enticed into listening to the explanations if they are well diluted with experiments. I also find that with very wide age mixtures, especially if there are quite a few younger children present, experiments involving the children are sure to hold the attention of the adults. For example in a recent tour abroad where this kind of mixture occurred on several occasions I found the following simple demonstration popular with both children and adults.

Demonstration 3.19

About six or eight children are asked to come out and stand in line all facing the same way and to close up one behind the other. Having discovered their names I then ask what will happen if I push John, the rearmost child, forward. Of course the answer is that Nicki, the girl at the front will fall over. But who pushed Nicki?—and gradually we arrive at the answer that although I did not push Nicki directly, my 'push' travelled along the row. And that of course is quite a good model of a compression wave. I always enjoy watching the smiles on the faces of the adults while this is going on, and it is obvious that it illustrates an important scientific idea in a painless and interesting way.

The second level that needs to be maintained is the explanation or elaboration for the non-scientific, but interested and intelligent adults. For example I might follow the 'child-wave' demonstration with a wave model, such as the Vinycomb model (see figure 3.19).

And the third level is the occasional tit-bit of scientific information or allusion to keep the attention of the specialists in the audience. For instance in doing the demonstration with the lensless slide projector described above (Demonstration 3.10) I might very well say at one point that the scientists present will recognise this as a way of presenting Fourier transform ideas. If done in this way, a little scientific jargon does no harm to the rest of the audience. Indeed it may do some good. Mendoza[71] made the excellent point that one of the causes of the mental block that so often bedevils people coming into science is the unfamilarity of the language. But if they have heard the terms earlier on without necessarily understanding them and before actually needing to use them, there is a familiarity that can be very helpful. For this reason he suggests that words like 'entropy', 'wavelength', etc, should be mentioned in passing by teachers long before the time comes for the pupils to study the related ideas seriously.

If all three levels can be kept going at the same time then, in my experience, the mixed ages and experience of the audience do not matter.

3.9 AUDIENCE PARTICIPATION

3.9.1 Different ways of participating

Audiences love joining in, but the type of participation and the ease with which it can be achieved varies greatly with age. In the 6–11 age group a call for volunteers will usually cause problems in keeping the number down rather than of shortage. There are four main ways of involving the audience at this level. The first is that mentioned above in

Demonstration 3.17—the use of questions to the audience in order to maintain their interest—and figure 3.24 illustrates a typical response.

Figure 3.24 Typical response of a class of 7–11 year-olds to a question. (Reproduced by permission of the *South Wales Echo*.)

3.9.2 Playing games

The second type of participation involves playing a kind of game, which is perhaps best described as an animated visual aid. My example is one that I use in talking about the origins of colour in everyday objects.

Demonstration 3.20

The aim is to illustrate the phenomenon of selective absorption. One volunteer is asked (without the audience hearing) to pretend to be the dye in a magenta-coloured flower that absorbs green and hence to grab any green card held by anyone passing by. A queue of maybe ten or fifteen children who have been given (randomly) either a red, blue or green card then file past and afterwards are asked to hold up their cards. Obviously, in this case, the greens have been removed and the mixture of blue and red cards left will mix to give the sensation of magenta. Of course this can be repeated with objects that absorb two of the primaries. And finally the volunteer is told to take all cards and the audience is left to deduce that the 'colour' represented must be black.

3.9.3 *Experiments with children*

The third way of involving children is to use them as an integral part of an experiment. Figure 3.25 shows two examples of this approach.

Demonstration 3.21

It is quite difficult to discuss the idea of momentum explicitly with young audiences and yet, if we wish to explain why we wear safety belts in cars and why cars need a 'crumple zone' at the front and back in order to minimize injury some way of getting round the difficulty must be found. In figure 3.25(a) I had invited a member of the audience on the platform and asked her to stand in a cardboard box. I then asked if she would mind having 2 kg of dried peas poured on to her feet and, though puzzled, she had no objection. I then produced 2 kg weight from behind my back and both her and the audience's reaction made the point for me with little further development; a lot of little collisions are far less painful than one big one!

Demonstration 3.22

Why do some materials conduct heat well and others hardly at all? My illustration uses a row of children (figure 3.25(b)) holding hands, the child at one end is asked to swing the arms back and forth and the oscillation travels down the row until they are all swinging. Then I stand in the middle and refuse to oscillate thereby giving a rough idea of what an insulator does.

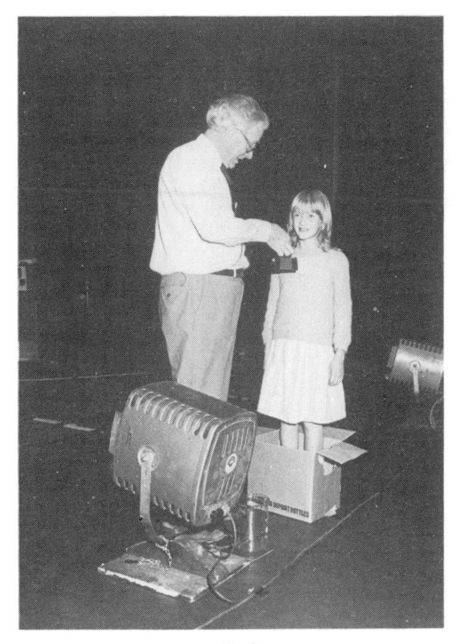

(*a*)

(*b*)

(*c*)

Figure 3.25 (*a*) Do you mind if I drop this on your toes? (*b*) Line of children demonstrating conduction. (*c*) The effect of an insulator. (Photos by R S Watkins.)

3.9.4 Help from the audience

The fourth type of audience involvement with young children is to get them to help with an experiment. In figure 3.26 a member of the audience is playing a recorder and observing the steady waveform on the oscilloscope, and of course there are many other ways in which children can help. The age group 12–16 is more difficult. They are less likely to answer questions from the floor and only a few extroverts will actually be prepared to come and assist spontaneously. However, if asked beforehand it is usually possible to find people to help with demonstrations of the third kind mentioned above, i.e. those in which help is needed in performing an experiment with inanimate apparatus. Grown ups are usually quite happy to help, especially if asked beforehand. But the best way of involving an adult audience is experimentation *en masse*. For example in the 'brain-washing' experiment described in Demonstration 2.7 (p. 77) or in Demonstration 3.4 (p. 114) on the three types of colour receptor in the eye.

Figure 3.26 Playing a recorder to study its waveform on the oscilloscope. (Photo by R S Watkins.)

3.10 SAFETY

3.10.1 Introduction

It should go without saying that safety in performing lecture demonstrations, both for the audience and for the lecturer and the assistants, is of paramount importance but, for completeness, I would like to say a few words about the kind of considerations that should apply.

We have become much more safety conscious in the last two decades or so and it is extraordinary to look back at some of the risks that were taken by previous generations. For example, William Spottiswoode in 1878 gave an evening discourse at the Royal Institution entitled 'A Nocturne in Black and Yellow'. The topic was optical interference with long path differences where, in white light, very pale colours are seen (for example in the outermost circles of Newton's rings). If sodium light is used then the fringes extend very much further, and of course are coloured black and yellow—hence the title. Spottiswoode needed a bright source of light in order to project the fringes on to a screen so that they could be seen by a large audience and subsequently[72] wrote

> For a burner adapted to lecture purposes I am indebted to Professor Dewar. The burner consists of an oxy-hydrogen jet, with the addition to the hydrogen tube of a chamber containing metallic sodium. The metal is volatized by a Bunsen's burner placed below it; so that the hydrogen emerges charged with sodium vapour. The result is a bright monochromatic light.

The safety implications of this demonstration, done in front of a distinguished audience of more than 450 people, make one gasp!

But, quite apart from such obviously hazardous experiments, there are quite a number of relatively simple ways in which an audience can be put at risk. Perhaps one of the most obvious, but nevertheless common, sources of danger is that of electrical connections of various kinds. Loose cables which might cross the path of members of the audience should be secured to the floor with adhesive tape; care should be taken that extension leads carry satisfactory earths and are correctly wired and that the load bearing capacity of the cable itself is adequate. Obviously the total load placed on any outlet socket must not exceed its rated value, but when using multi-socket distribution boards this is a point that can easily be overlooked.

3.10.2 Teaching by example

Quite apart from the obvious need to ensure the safety of the audience and of the lecturer, the demonstration lecture has great potential for teaching good safety practice by example. Safety glasses should be worn whenever there is the slightest risk to the lecturer's eyes; a safety glass screen should be placed between the lecturer and the audience if there is the remotest risk of even a tiny explosion; a fireproof board should be placed on the bench if any experiments involving heating are to be performed; a large tray (e.g. a photographic dish) should be placed under any apparatus where even a mildly corrosive or staining liquid might be spilled. It is not my intention here to provide an exhaustive list of safety precautions, but to draw attention to the enormous potential for implicit safety education in lecture demonstrations; audiences, especially young ones, absorb a great deal more than just the topic being discussed and this powerful influence must not be ignored.

Epilogue

I have covered a very wide variety of topics, perhaps not always in the most logical order but I hope, if nothing else, that I have conveyed a sense of my own enthusiasm for the lecture demonstration as a means of communication. It seems to work with all age groups and is a great way of inculcating a sense of excitement about science, especially in children.

But will the art and science of lecture demonstration survive? In 1970 Gerald Holton contributed an article to the classic volumes on lecture demonstration produced by the American Association of Physics Teachers[73] that began with the question 'Is the human demonstrator obsolete' and I should like to quote one or two extracts from his answer. He begins by discussing some disadvantages of lecture demonstration (to which I made reference in the Prologue) and then goes on to reflect on the reasons why demonstration is becoming more difficult and cites two reasons:

> One lies in the nature of the advance of physical science. We have long passed the time when, by means of pulleys, magnets, and electroscopes one could demonstrate the current state of science—and consequently when occasionally a fundamental discovery could be made in the course of preparing or giving an elementary lecture, as was the case with Kepler, Victor Meyer, Oersted, and Hertz[74]. . . .
>
> . . . A second reason for the increasing difficulty with demonstrations lies in the practical realm. While physics classes have been growing and multiplying, new physics lecturers are, more and more often, men who have passed through no substantial period of teaching apprenticeship.

> They usually face a large class for the first time when they are put fully in charge of a course. Nor can they count on skillful and reliable curators to help behind the scene. . . .

He then asks whether we may now be moving into the era of the 'automation of teaching' and the 'technological classroom' and quotes a remark made by Eric Rogers[75] in 1959:

> So I look forward to good physics teaching going on fifty years from now, but mostly by film and TV, from sheer demand of numbers.

Holton then goes on to make what seems to me to be a very significant point:

> There is no question but that all bad lecture demonstrations (and lecturers) should be replaced, if necessary, by good audio-visual surrogates, just as we should let much of the rote teaching be done by means of programmed devices. . . .
>
> . . . But we must not get caught up in our usual optimism that any new, promising development should be extended as quickly as possible from horizon to horizon. . .
>
> . . . The human demonstrator is obsolete only if we compare the few good films, or the rare good use of TV with a mediocre or bad lecturer; the majority of currently available audio-visual aids is not of high quality, and I fear this will remain true for a long time. . .
>
> . . . And above all, there are a number of important values which, it appears to me, are conveyed in the interaction between the live lecturer and the student in the same room, particularly through the scientific demonstration, values which depend on communication being initiated by an actual person rather than by a surrogate.

I commend the rest of this article, and indeed much of the other material in the introductory section of these valuable volumes, to any one who wishes to read more about the subject than it has been possible to cover in this small book.

But I feel that the last sentence in my quotation from Holton is of supreme importance. I firmly believe in the future of demonstration, though obviously its popularity will rise and fall as time goes by. Twenty years ago we were told that the theatre was on the way out and was being killed first by the cinema and later by television. Yet now the live theatre seems healthier than ever. There really is nothing quite like live theatre or live music. I remember visiting an opera house with a friend who was not often able to hear a live performance but spent a great deal of time listening to records on his stereo hi-fi system. As soon as the orchestra started playing he whispered in my ear 'Isn't the stereo fantastic!'

I am sure that the attitudes to live lecture demonstration are likely to be the same. As Holton says, film or video recording is better than nothing but there is no substitute for the interaction between a good lecturer and the audience. I remember seeing a cartoon many years ago when the tape recorder was first introduced as a teaching tool in universities, though regrettably I cannot remember where I saw it or who drew it. It showed a large tape recorder at the front of an empty lecture room with a small tape recorder on every desk listening to it! It made the point beautifully and reminded me of the rather sour definition of a lecture that used to be current in my student days 'A lecture is the means by which the notes of the lecturer become the notes of the student without passing through the minds of either.'

I hardly think that definition could apply to a well delivered demonstration lecture. In fact the one major criticism I have sometimes heard of demonstration lectures in undergraduate courses is that it is difficult to get a good set of notes. I suspect that such a criticism really implies that the person making it is not sufficiently well trained as an observant scientist and was perhaps hoping to have notes dictated.

A great deal of emphasis is being placed on informal science education at the moment. By informal science education I mean science education that is not geared to certificates or examinations or degrees. Science and technology play an important part in all our lives and it is vital that all citizens should at least know something about it; there is therefore great scope for the media, and particularly television, to help in giving science a better image than it has tended to have in recent years. Obviously demonstration lectures for the general public and for various levels of children at school have an important part to play.

Ten or fifteen years ago it was not considered to be part of the duty of a professional scientist to take part in such activity. In fact some members of the scientific establishment tended to feel that for a research scientist to 'waste' time in lecturing to schools or to the public was not really respectable. But thankfully, as I said in the Prologue, the pendulum has begun to swing the other way. I referred there to the Royal Society's report 'The public understanding of science'[76] and I should like to quote two passages from the summary:

> Science and Technology play a major role in most aspects of our daily lives, both at home and at work. Our industry and thus our national prosperity depend on them. Almost all public policy issues have scientific or technological implications. Everybody, therefore, needs some understanding of science, its accomplishments and its limitations...
>
> ... Scientists must learn to communicate with the public, be willing to do so, and indeed consider it their duty to do so. All scientists need, therefore, to learn about the media and their constraints and learn how to explain science simply, without jargon and without being condescending. Each sector of the scientific community should consider, for example, providing training on communication and greater understanding of the media, arranging non-specialist lectures and demonstrations, organizing scientific competitions for younger people, providing briefings for journalists and generally improving their public relations.

This recognition by the premier scientific body in Britain marks an important step forward and I hope that established scientists will take up the challenge. Perhaps this monograph may be of some small help in furthering the cause of increasing public understanding of science.

References

1. The Royal Institution 1974 *Advice to Lecturers* (London: Mansell) p. 19
2. Bishop G D 1961 *Physics Teaching in England from Early Times up to 1850* (London: PRM publishers)
3. Desaguliers J T 1763 *Course of Experimental Philosophy* preface
4. Yorke P quoted in Winstanley D A 1935 *Unreformed Cambridge* (Cambridge: Cambridge University Press) p. 151
5. Pepper J H 1860 *The Boy's Playbook of Science* (London: Routledge)
6. Caroe G 1985 *The Royal Institution* (London: John Murray) p. 129
7. Galloway R 1881 *Education, Scientific and Technical* (London: Trübner) p. 53
8. The Royal Institution 1974 *Advice to Lecturers* (Mansell) p. 5
9. The Royal Society 1985 *The Public Understanding of Science* (London: The Royal Society)
10. Tyndall J *Lectures on Sound 1854–82* (manuscript notes in the archives of the Royal Institution)
11. Desaguliers J T 1763 *Course of Experimental Philosophy*
12. Gregory R 1986 *Hands-on Science* (London: Duckworth)
13. The Chemist & Druggist 1891 *Scientific Mysteries* p. 73
14. Hauksbee F 1709 *Physico-Mechanical Experiments on Various Subjects*
15. Pohl R W 1970 in *Physics Demonstration Experiments* ed H F Meiners (American Association of Physics Teachers) chapter 3
16. The Chemist & Druggist 1891 *Scientific Mysteries* p. 97
17. Merrill F H and von Hippel A 1939 *J. Appl. Phys.* **10** 873
18. Bishop G D 1961 *Physics Teaching in England from Early Times up to 1850* (London: PRM publishers)

19 Tyndall J 1868 *Heat; a Mode of Motion* 3rd edn (London: Longman) p. 251
20 Tyndall J 1898 *Heat; a Mode of Motion* 11th edn (London: Longman) p. 205
21 Tyndall J 1882 *Six Lectures on Light* 3rd edn (London: Longman) p. 143
22 Tyndall J *Lectures on Sound 1854–82* (manuscript notes in the archives of the Royal Institution)
23 Tyndall J 1898 *Heat; a Mode of Motion* 11th edn (London: Longman) p. 273
24 Tyndall J 1875 *Sound* 3rd edn (London: Longman) p. 105
25 Wylde J *c.* 1860 *Magic of Science* (London: Griffin) p. 238
26 Helmholtz H L F 1885 *On the Sensations of Tone* 2nd English edn (New York: Dover, 1954)
27 Galloway R 1881 *Education, Scientific and Technical* (London: Trübner) p. 279
28 Galloway R 1881 *Education, Scientific and Technical* (London: Trübner) p. 280
29 Brock W H 1973 *H. E. Armstrong and the Teaching of Science, 1880–1930*
30 Jenkins F W 1979 *From Armstrong to Nuffield* (Edinburgh: John Murray) p. 91
31 Bragg Sir Lawrence 1954 Models of metal structure *Proc. R. Inst.* 844
32 Miller J Sumner 1966 *Millergrams* (Ure Smith)
33 Caroe G 1985 *The Royal Institution* (Edinburgh: John Murray) p. 45
34 Tyndall J 1857 *Proc. R. Inst.*
35 Perry, Lord 1984 *The State of Distance Learning World Wide* (report published by International Centre for distance learning of the United Nations University)
36 Pizzey S 1987 *Interactive Science and Technology Centres* (London: Science Projects Publishing)
37 Taylor C A 1987 Dramatic events in science education *Phys. Ed.* **22** 294
38 Bell J 1983 How to pack a public lecture *New Scientist* **100** 584
39 Beetlestone J and Taylor C A 1982 Linking science and drama at school *Impact of Science on Society* **32** 473
40 Taylor C A 1987 *Diffraction* (Bristol: Adam Hilger) p. 36
41 Spottiswoode W 1876 *Polarisation of Light* (London: Macmillan) p. 36
42 Taylor C A and Lipson H 1964 *Optical Transforms* (Bell)

43 Welberry T R 1985 Diffuse x-ray scattering and models of disorder *Rep. Prog. Phys.* **48** 1543
44 Taylor C A 1978 *Images* (Wykeham) p. 81
45 Taylor C A 1985 Images *Proc. R. Inst.* **57** 238
46 Attenborough D 1978 *Life on Earth* (London: BBC Publications)
47 Taylor C A 1987 in *Science Education and Information Transfer* (Oxford: Pergamon) p. 13
48 Tickton S G *To Improve Learning: an Evaluation of Educational Technology* (London: Bowker)
49 Postlethwaite S N and Mercer F V 1974 *New Trends in the Utilization of Educational Technology for Science* (Unesco) p. 186
50 Boulton M 1987 in *Science Education and Information Transfer* (Oxford: Pergamon) p. 54
51 Sumner G 1985 *J. Geog. in Higher Education* **8**
52 Stonier T 1987 in *Science Education and Information Transfer* (Oxford: Pergamon) p. 212
53 The Royal Institution 1974 *Advice to Lecturers* (Mansell) p. 6
54 The Royal Institution 1974 *Advice to Lecturers* (Mansell) p. 21
55 Taylor C A 1981 in *John Tyndall, Essays on a Natural Philosopher* (Dublin: Royal Dublin Society) p. 211
56 Kenny P 1982 *A Handbook of Public Speaking for Scientists and Engineers* (Bristol: Adam Hilger)
57 The Royal Institution 1974 *Advice to Lecturers* (Mansell) p. 19
58 Barlex D and Carré C 1985 *Visual Communication in Science* (Cambridge University Press)
59 The Royal Institution 1974 *Advice to Lecturers* (Mansell) p. 20
60 Tyndall J 1882 *Six Lectures on Light* 3rd edn (London: Longman) p. 2
61 'A mere Phantom' 1870 *The Magic Lantern*
62 Tyndall J 1875 *Sound* 3rd edn (London: Longman) p. 168
63 Newton, Sir Isaac 1730 *Opticks* (New York: Dover, 1952) Book 1, Proposition V
64 Land E 1977 *Scientific American* (December) 100
65 The Royal Institution 1974 *Advice to Lecturers* (Mansell) p. 17
66 Taylor C A 1978 *Images* (Wykeham) p. 2
67 Taylor C A 1987 *Diffraction* (Bristol: Adam Hilger) p. 20
68 Foxcroft G E 1975 *Proc. R. Inst.* **48** 166

69 Vinycomb T B, described by Alexander Wood in his Christmas Lectures for 1928 and published as *Sound Waves and their Uses* (Glasgow: Blackie, 1930)
70 Barton E H 1918 *Phil. Mag.* **36** 169
71 Mendoza E 1965 Waves *Contemp. Phys.* **6** 217
72 Spottiswoode W 1878 A nocturne in black and yellow *Proc. R. Inst.* 582
73 Holton G E 1970 in *Physics Demonstration Experiments* ed H F Meiners (American Association of Physics Teachers)
74 Ramsauer C 1953 *Grundversuche der Physik in Historischer Darstellung* (Berlin: Springer)
75 Rogers E M 1959 *Proc. Wesleyan Conf. on Lecture Demonstrations* p. 6
76 The Royal Society 1985 *The Public Understanding of Science* (London: The Royal Society) p. 6

Index of Demonstrations

Categories (as defined in §2.1.1) are indicated by the codes V for category 1 (Visual aids using non-conventional apparatus), A for category 2 (Analogue demonstrations) and R for category 3 (Real experiments). Demonstration numbers are as in text.

Index